CAMERAS IN THE CLASSROOM
Educating the Post-TV Generation

MICHAEL SCHOONMAKER

ROWMAN & LITTLEFIELD EDUCATION
Lanham, Maryland • Toronto • Plymouth, UK

Published in the United States of America
by Rowman & Littlefield Education
A Division of Rowman & Littlefield Publishers, Inc.
A wholly owned subsidiary of The Rowman & Littlefield Publishing Group, Inc.
4501 Forbes Boulevard, Suite 200, Lanham, Maryland 20706
www.rowmaneducation.com

Estover Road
Plymouth PL6 7PY
United Kingdom

Copyright © 2007 by Michael Schoonmaker

All rights reserved. No part of this publication may be reproduced, stored in a retrieval system, or transmitted in any form or by any means, electronic, mechanical, photocopying, recording, or otherwise, without the prior permission of the publisher.

British Library Cataloguing in Publication Information Available

Library of Congress Cataloging-in-Publication Data

Schoonmaker, Michael.
    Cameras in the classroom : educating the post-TV generation / Michael Schoonmaker.
       p. cm.
    Includes bibliographical references.
    ISBN-13: 978-1-57886-505-5 (hardcover : alk. paper)
    ISBN-10: 1-57886-505-0 (hardcover : alk. paper)
    ISBN-13: 978-1-57886-506-2 (pbk. : alk. paper)
    ISBN-10: 1-57886-506-9 (pbk. : alk. paper)
    1. Television cameras. 2. Television—Production and direction. I. Title.
  TR882.5.S35 2006
  372.67'2—dc22   2006023826

∞™ The paper used in this publication meets the minimum requirements of American National Standard for Information Sciences—Permanence of Paper for Printed Library Materials, ANSI/NISO Z39.48-1992.
Manufactured in the United States of America.

Contents

Prologue		v
Introduction: Unlocking the Moviemaking Mind		1
1	Children of the TV Generation	15
2	Reading, Writing, and Television	30
3	Voice of the Developing Mind	47
4	The Message Is the Message: Media Literacy in a Visual Age	64
5	Writing with Cameras	78
6	Bonding with Cameras	93
7	Videomaking and Curricula	109
8	Video Playing Field	122
9	Thinking in Video	134
10	Videoplay: Tapping into the K–12 Imagination	149
11	Inventive Video	163
Epilogue		175
References		180
Photo Credits		186

Prologue

"**If there** were only a magic wire to connect our visual imagination to a movie screen, we'd be rich beyond our wildest dreams": This is what I tell students in my beginning production classes as they screen—usually with some level of despair—their "first cuts." What they have managed to convey on the screen pales in comparison to the mind's-eye vision that inspired their moviemaking efforts.

If there were such a "magic wire," it could transport their dreams, memories, imaginative ideas, personalities, passions, the stories of their lives, and much, much more straight to the screen. But until such a wire is discovered, real-life moviemaking is likely to remain a painstaking challenge however easy it is to dream it up in our heads.

Interestingly enough, the difficulty of the process rarely discourages a film-maker who has a story to tell. They persist because there is something more powerful that inspires them: the immense fulfillment in sharing their visions with others. The notion of touching other hearts and souls seems to inflate a filmmaker's self-esteem and sense of belonging in the world outside of himself or herself.

Over the years I have found this sharing spirit to be one of the most potent forces behind successful movies and television. Imagine my surprise when I discovered the same phenomenon in a group of kindergarteners at Corrigan Elementary School. But, in this case, sharing the movies in their minds had nothing to do with the motion picture industry. It was more about being human in a very visual world.

INTRODUCTION
Unlocking the Moviemaking Mind

CAMERAS *in the Classroom* began when my son's kindergarten teacher allowed me into her class to give a show-and-tell lesson about how television is made. It didn't take long to discover that Mrs. Brighton's kids were far more interested in actually *making* television than talking about it or even watching it.

When I thought about it, I realized my little lesson was the equivalent of teaching children how to read what other people wrote without considering whether they might want to write themselves. So I asked Mrs. Brighton if she would be okay with the idea of letting the children play around with a video camera to see what we could create together.

At first she was apprehensive because she had no idea how to make a movie, let alone whether it would be a worthy classroom activity. Like all teachers, she had much to accomplish in her curricula and limited time to get it done. When I assured her that I would be with her the whole time and take care of all the technical details, she agreed that it would at least be a fun experiment.

She explained to the children that we were going to make a "video play." After talking with her class assistant, Mrs. DeBella, she asked if we might be able to work the experience into her weekly alphabet lesson. The students spent an entire week on each letter of the alphabet, and this week's letter was *i*.

It was decided we would frame the movie around the *i* word *invisible*. This decision marked the beginning of what has become a noteworthy synergy between making television and public school curricula.

INVISIBLE JUICE

<u>INTERIOR—MRS. BRIGHTON'S KINDERGARTEN CLASSROOM—MORNING</u>

THIS PARTICULAR DAY BEGINS LIKE MOST OTHER SCHOOL DAYS: MORNING ANNOUNCEMENTS, ATTENDANCE, READING, AND LEARNING GROUPS WITH THE CLASS ASSISTANT, MRS. DEBELLA. NOW IT IS TIME TO FORM THE CIRCLE FOR ONE OF THE MOST ANTICIPATED ACTIVITIES OF THE DAY—SNACK TIME.

                  MRS. BRIGHTON
All right, boys and girls, who brought snack today?

                  TROY
   (Unable to contain his restlessness)
                  I did!

                  MRS. BRIGHTON
      (with a gentle smile)
What did you bring for the class, Troy?

                  TROY
        Juice! *Invisible* juice!

MRS. BRIGHTON AND MRS. DEBELLA SEARCH EACH OTHER FOR ANY EXPLANATION OF THIS RATHER ODD IDEA. FINDING NONE, THEY TILT THEIR HEADS IN TROY'S DIRECTION.

                  MRS. BRIGHTON
            Invisible Juice?

                  TROY
               Yes!

TROY RAISES AN EMPTY TRAY NORMALLY RESERVED FOR BEVERAGE CUPS.

TROY
              Does anyone want some?

THE CLASS SHOUTS OUT BEFORE MRS. BRIGHTON CAN SAY
ANYTHING.

                    CLASS
                    Yeah!

MRS. BRIGHTON AND MRS. DEBELLA RETURN CONFUSED
GLANCES, THEN SHRUG IN RESIGNATION. TROY CAREFULLY
HOLDS THE TRAY OUT TO EACH OF HIS CLASSMATES UNTIL ALL
HAVE COLLECTED THEIR INVISIBLE PORTIONS. THEN, RAIS-
ING THE LAST INVISIBLE CUP TO HIS LIPS HE ANNOUNCES:

                    TROY
                  Drink up!

THE REST OF THE CHILDREN FOLLOW HIS LEAD.

                    TROY
               Now let's count

THE OTHER CHILDREN JOIN IN.

                    CLASS
             One, two, three . . ."

ON "THREE" THE CHILDREN CLAP, THEN DISAPPEAR
INSTANTLY. MRS. BRIGHTON AND MRS. DEBELLA LOOK ALL
AROUND THE ROOM, THEN AT EACH OTHER IN AMAZEMENT.

                  MRS. DEBELLA
      Well, I guess it really was invisible juice!

JUST AS MRS. BRIGHTON NODS IN AGREEMENT, THE INVISI-
BLE CHILDREN—WHO HAVEN'T LEFT THE ROOM DESPITE THEIR
INVISIBLE STATE—BURST OUT IN ENORMOUSLY INVISIBLE
LAUGHTER.

                   THE END

Making Television with Children

Invisible Juice was a big hit with children and the teachers. The "visual conspiracy" of using a video camera to make the class magically disappear was being connected to learning activities of the class. At the same time, they were learning something about the media. They watched their little movie over and over and demonstrated great pride in what they saw as their collective creation—not to mention the fact that *i* was now an especially memorable letter of the alphabet.

Both Mrs. Brighton and Mrs. DeBella caught the moviemaking bug. They came up with ideas for several more video activities—some with connections to lesson plans, others for the sake of documenting the class experiences. *Invisible Juice* also developed two sequels in following years.

I also caught the bug and continued making movies with kids as my three sons progressed through the grades. Recently, I found myself reminiscing with fifth-grade teacher Mary Catherine Catrell about the movies we did in her class.

> The first time that you came in, I thought, "How are we gonna do this?" and, "I don't know how to do this." And what happens is the kids know how to do this! And they love it.... It's like a different venue for them.

She was referring to the music video we made to complement her poetry unit. As with the other class videos we had made up to this point, blending videomaking with curricular activities was proving a resounding success.

The kids in Mary Catherine's class were intuitive and surprisingly skilled with visual media—both the technology and technique. But more importantly, they got more involved with poetry than they would have through more traditional instruction. Mary Catherine was finding that videomaking truly invigorated her poetry lesson plan.

Making television was without a doubt as positive an experience for her class as it had been for every other class up to that point. I began asking myself, "Why do these positive things happen so consistently? And how is it that children, in Mary Catherine's words, 'know how to do this' even though they have not been trained in videomaking technology and procedures? And why are the children so at ease with visual expression as a learning venue? Can these K–12 videomaking experiences be effectively applied to other school settings and curricula?"

It was at this point that I set out to answer these questions by exploring what happens when videomaking tools—a video camera, video-editing software, and some storytelling conventions—are applied to K–12 lesson plans.

To date, the teachers and I have made more than thirty movies involving kids from ages five to eighteen. The movies range in style and genre from short movie stories, information/news/history programs, animation/illusion/special effects programs, music videos, documentations of field trips, athletic competitions, story sequels, and television series.

It is through these experiences with children, teachers, lesson plans, and videomaking tools that the dynamics of visually enhanced learning are explored. Think about it: television that's good for kids; television that stimulates learning. This is not just any television, like the kind kids watch when they just need to veg out or escape. This is television children make themselves.

There is much to share from these experiences and much to apply to the ever-evolving challenges K–12 educators face today. But before I get to that, I'll address some more basic questions about the book.

What Is the Book About?

Cameras in the Classroom brings together two powerful icons of our time: primary education and moving-image media.

It is about a new and slowly unfolding educational discourse: visual thinking and expression in K–12 learning environments. It is based on fourteen years of videomaking with children and teachers from all levels of the public school spectrum, in two large city school districts (100,000+ students each), one in New York and one in Pennsylvania. It is important to realize that findings are derived from the good, bad, and ugly realities of everyday classrooms, amidst the sometimes tumultuous and backbreaking challenges these systems face. These experiences are not sanitized and they're not contingent on privileged resources.

Readers will not only see how video can be used as a learning tool but, more importantly, *why* video works as well as it does in ordinary educational settings. Such an understanding will enable them to apply and evolve the lessons of the book to their unique needs, whether connected to teaching or to media literacy research and theory.

This is not a "how-to" book as much as it is a "why-to" book: why to rethink the relationship of television and children as television has become increasingly available to them as creators, not just consumers. In this light, *Cameras in the Classroom* is a collection of experiences and discoveries that can be used as a base from which to generate more discoveries.

I receive regular inquiries from K–12 teachers and even colleagues in other disciplines (history, anthropology, law, cultural studies, special education, political science, textual) asking how they might apply videomaking to particular learning objectives. They are usually interested in some sort of one-course videomaking workshop that could help bolster their confidence and give them some ideas on how to get started. *Cameras in the Classroom* will provide that and more—the "more" being a contextual understanding of videomaking as a means of expression that can effectively serve a wide variety of agendas—professional, intellectual, or organic.

Who Is This Book Written For?

For those interested in media related to education—students, parents, teachers, researchers, and producers—this book will offer fresh and exciting findings from kindergarten to twelfth-grade classrooms that can be applied to educational environments and promote further research and exploration.

Primary and secondary school teachers will find *Cameras in the Classroom* approachable, stimulating, and relevant to a wide variety of learning environments.

Researchers and administrators involved in educational reform will find valuable substance in the real classroom accounts of what happens when videomaking is applied to traditional learning activities.

Concerned parents will find insights on tackling one of life's tougher assignments: raising children in a very visual world. Given the volume of visual media—TV, home video, MP3 players, DVD, video games, and the Internet—the parents I have talked with often find themselves on the defensive, asking "How much is *too much?*" There is no simple answer to this question, but this book can help them see a refreshingly different application of visual media to their children's lives.

Who Is This Book Written By?

Cameras in the Classroom is written by a parent, a producer, and a very curious teacher—all in one.

The Parent

As a father of three boys, I involved myself in their classes at every opportunity because I wanted to be part of their education. The more time I spent in the classroom, the more opportunities I found to help other children too.

This is what led me to bring my video camera into the classroom—at first as a show-and-tell piece, but as time went on as an educational tool. I believed I could teach kids about the magic of television and filmmaking, as an "experienced professional." After all, I'd been involved in some aspect of media making for most of my adult life: first as a production professional at MTV and NBC, and then as a professor in the Television-Radio-Film Program at Syracuse University.

However, the more I got involved with the classes, the more I realized how little I knew about the everyday challenges of teaching children. I had assumed the learning process was much easier than it actually was. Inside my pristine notion of the modern schoolhouse, there were no state mandates or racial inequities or severe imbalances between learners. The classrooms I worked in had all these challenges *in addition to* the challenge of delivering a structured education.

In the end, I learned much more than I could ever have taught—about children, teachers, schools, and the learning process. The closer I worked with the children, the more I could relate to their perspectives of media, particularly if I took the time to listen to them instead of just filling their heads with my "wisdom."

The Producer

When I walked into that first kindergarten class, the "television producer" in me wanted to figure out how moving-image media could work better for children in the classroom. No matter how much I loved television and cinematic entertainment, they had very bad "raps" in educational circles. Most of the parents and teachers I have known perceive television as a threat to education. Parents assume that watching television makes their kids lazy and dulls their motivation to learn. Teachers tend to assume that television spoon-feeds information and saturates it with entertainment and special effects that they cannot compete with. The dreaded "short attention span," archenemy of education, was to these people a result of children being brought up on television.

Earlier in my career as a production manager at MTV, I had learned to walk a line between reactions to the work that I did. Acquaintances my age and younger thought I had the "coolest" job, naturally because they were big fans of the channel. But most parents I talked to kept my conscience in check. They tended to be (politely) critical of the channel's irreverent projections of life and my subsequent involvement with such activities.

I had always liked what I did and felt proud of my role in touching young lives—but I could never completely push aside the possibility there might be some harm, however invisible to me, to young viewers. As someone who knew about the people and process behind the scenes of the industry of media making, I had always had a gut feeling my chosen field wasn't as bad as the pundits, like Neil Postman, claimed,

> When a population becomes distracted by trivia, when cultural life is redefined as a perpetual round of entertainments, when serious public conversation becomes a form of baby-talk, when, in short, a people become an audience and their public business a vaudeville act, then a nation finds itself at risk; culture-death is a clear possibility. (1985)

As a media professional I was involved in a business-driven venture, but what I did in my view was much deeper than "baby-talk" or a vaudeville act with no conscience. At MTV, I worked with caring parents, active and intelligent citizens, and imaginative artists. What I did as a professional producer of media was not beyond reproach, but neither was it was devoid of social value.

In fact, one of the most important lessons I ever learned as a producer, and have tried to pass on to my students, is the tremendous responsibility all mediamakers carry as storytellers. Viewers look to the stories we tell to make sense of their lives. The more seriously we take this responsibility, the greater the impact our stories will have on people. There is no more rewarding outcome to the TV producers and filmmakers I know than to have a profoundly positive impact on other people. It is this creative spirit behind the process of television and filmmaking that tends to go unnoticed when it comes to protecting young minds from the "possible" dangers of television.

The Teacher

As a researcher interested in the complexities behind "seemingly simple" television viewing, I found it odd that I knew so little about children and their viewing behaviors and perspectives. In past research (Schoonmaker 1994), I studied how television viewers chose home videos in video stores. Even though children were part of the study, I learned little to nothing about them and what they thought of video or television—in part because they were so insulated by their parents.

I never had any problem finding out what adults thought of children and media. It was what the children *themselves* thought of media that I had trouble finding. And in the limited interactions I had with children, I found it particularly challenging to get anything out of them—they simply didn't respond well in short, "adult-centered" interviews.

The biggest question I had when I first started making video projects with my son's kindergarten class was, "What does television mean to these kids and how do they use it in their lives?"

But I also carried an intense optimism that television was capable of contributing more to education than it had since its inception more than a half century ago. It was around that time that journalist Edward R. Murrow noted the same:

> This instrument can teach, it can illuminate; yes, it can even inspire. But it can do so only to the extent that humans are determined to use it to those ends. Otherwise it is merely wires and lights in a box. There is a great and perhaps decisive battle to be fought against ignorance, intolerance and indifference. This weapon of television could be useful. (1958)

Once I started making movies with the kids, it didn't take long to find the answers to my questions. Interaction through experience provided more insights into children's perspectives than I could have hoped for with a sit-down interview. It was one thing to talk to children about their perspectives and ideas, and quite another to participate directly with them as they created and experienced the process of making television. The more time I spent creating videos with kids and their teachers, the easier it was to understand them because I was doing it on their terms, not my own. It was at this point that I realized I could gather more than just their perspectives on media, which turned out to be fairly straightforward. In K–12 classrooms, I was in a position to observe the synergy between videomaking and learning.

How Is the Book Organized?

Cameras in the Classroom is organized along recurrent themes over eleven chapters that illustrate the dynamics of videomaking in learning environments:

- Videomaking, like other forms of media production, is very intuitive to children, even at very young ages.
- Videomaking inspires children to read.
- Videomaking helps children better understand themselves.
- Videomaking educates children about the media and helps them harness its power.
- Videomaking helps children better express themselves.
- Videomaking helps children connect with each other and their surroundings and to better experience the value of education.
- Videomaking stimulates learning and helps make subjects more relevant to children.
- Videomaking levels the playing field of learning.
- Videomaking helps children widen their learning landscape by activating their in-built visual sense.
- Videomaking helps channel a child's imagination into the learning process.
- Videomaking stimulates critical thinking.

Although these themes are dealt with individually in chapters, they do not operate in distinct vacuums. They are handled separately for the sake of clarity and explanation.

Each chapter sets out to illuminate its theme with

- experiences from K–12 videomaking activities
- content from creative works of children and teachers
- discussions with teachers about the videomaking in the context of their teaching environment
- explanation of videomaking procedures and their connection to learning activities and objectives
- relevant literature from the field that connects to videomaking experiences

How Does This Book Relate to Works of Others in the Field?

Although the intellectual underpinnings of *Cameras in the Classroom* are not exclusive to a single field of study, they are most closely rooted in media literacy, also referred to in European circles as media education.

There are different, sometimes opposing, schools of thought within the field of media literacy. *Cameras in the Classroom* identifies with what noted media education scholar David Buckingham termed a *Social Model* of television literacy.

The *Social Model* sees television literacy as a complex and creative process because children use television "as a means of negotiating social and cultural identities in quite diverse ways." Such an approach involves an open examination of television and the developing minds that use it to make sense of their worlds. Television literacy in this light is an intellectual resource not unlike reading, writing, and arithmetic: a tool of thought and expression, rather than the subject of a lesson plan.

The *Social Model* is in opposition to what Buckingham called an *Ideological Model*[1], which

> presumes that television literacy is a single set of cognitive abilities which individuals "possess," that meaning is objective and inherent in texts, and that both can be defined irrespective of social or cultural forces. Educationally, it gives rise to a narrow, mechanistic pedagogy which seems designed to produce conformity rather than genuine, open-minded critical enquiry. (1993a)

By contrast, the social model sees television literacy as a more complex and creative process because children use television "as a means of negotiating social and cultural identities in quite diverse ways." Such an approach involves a comparatively more open examination of television and the developing minds that use it to make sense of their worlds. Television literacy in this light is an intellectual resource not unlike reading, writing, and arithmetic: a tool of thought and expression, rather than the subject of a lesson plan.

From this social vantage point, *Cameras in the Classroom* informs and expands several questions and issues that have been unearthed by researchers in media literacy, including cross-disciplinary approaches, redefining educational television, and the relationship between media production and children.

Cross-Disciplinary Approaches to Media Literacy

The first question concerns the integration of media into traditional K–12 curricula. David Shaw, Pulitzer Prize winner and media/technology columnist for the *Los Angeles Times*, summarized this point nicely:

> Education in America is very structured, resistant to change, and Media Literacy is not an accepted course in the formal canon, not part of what is known as "K–12 standards." With school budgets tight everywhere, it's difficult to introduce programs or classes or hire more teachers to administer those programs and teach those classes. Media Literacy advocates say the solution is to incorporate in existing classes—cross-disciplinary approaches. (2003)

The book explores a wide variety of such cross-disciplinary activities from kindergarten ABCs to fifth-grade poetry to twelfth-grade business law. Incorporating media into existing subjects and teaching objectives is not only practical but also doubly effective: It teaches media literacy within the context of a visually enlivened lesson plan.

Redefining Educational Television in a Digital Age

Responding to "those who would rather blame the 'boob tube'" for America's educational ills—rather than the woefully underfunded school resources and teacher salaries, overcrowded classrooms, and lack of accountability of school systems—David Bianculli offers, "The truth is, if America really wants to shape up its educational system, a proper use of television and its new technologies may be the best and fastest means to that end" (2000). Part of this effort involves putting aside long-standing fears of television's negative effects on developing minds. This book demonstrates through example how harnessing television's strengths can invigorate lesson plans and learning environments.

Media Production and Children

Cameras in the Classroom answers Kathleen Tyner's call for a better understanding of optimum uses, potentials, and policies for hands-on student production:

> The only quantitative data currently tracked by educational bureaucracies about video in the schools is gleaned from sales figures about equipment purchases. Unfortunately, equipment availability is a poor indicator of what students are actually doing with video behind the closed door of the classroom.... Very little guidance is given to teachers about the optimum use of moving-image technologies. (1994)

This research is of growing importance as media-production technology becomes increasingly affordable and easy to use. This also complicates the notion of television as a one-way appliance—one way from TV programmer to mass audience. When consumers of television become producers of television, some very interesting things happen.

David Gauntlett's work with British schoolchildren confirmed this:

- The "labeled disadvantaged" seemed to genuinely thrive and in ways out perform the "non-disadvantaged" in unexpectedly positive and eloquent ways.
- The video project appeared to bring the subject of the environment "alive" for the children concerned.
- Children who have limited abilities in written English become able to demonstrate their creativity and intelligence.
- Children demonstrated a high level of media literacy in all age groups.
- The video-making process may have contributed to the sense of community-feeling which children had for their area. (1996)

Cameras in the Classroom expands upon these findings in the area of media production and unearths even more.

K–12 Education

Because my background is in communication and social science, I looked to the teachers I worked with as the experts and authorities on education. When I was confused about something that was happening in the way of educational dynamics, I consulted them, and they had loads of insights from both their educations and—perhaps most importantly—their teaching experiences.

> It's really learning to understand reality, and separating what you see on video and understanding that video allows you to make lots of things look very, very real. . . . I think for a lot of the kids it's so good to start to separate, because I think we're in an age where kids need to know that what they're seeing on television is not real.
> —A kindergarten teacher who had never made videos before

> When you're teaching kids in a setting where the economic structure and the home structure and the skill levels are so varied, some children—their world is very small. And I think any time you can expand that with outside experiences, you're expanding everything about them. You're giving them a new reference point, you're giving them new knowledge, you're giving them more critical thinking, new experiences they can relate to and I think it's really increasing their brainpower.
> —A first-grade teacher who had never made videos before

> Teachers spend an incredible amount of time trying to find the right medium to communicate a certain idea. . . . It's wonderful to be able to create the medium and then use it for my own purpose.
> —A second-grade teacher who had never made videos before

INTRODUCTION: UNLOCKING THE MOVIEMAKING MIND   13

You know the time machine and the magically appearing in all these different countries . . . I was like, are they going to get too far away from learning about other cultures, but when the movie was done and when I saw it I was like, "It worked!" It all came together really nicely, everybody learned something from it and they learned about social studies, about other cultures, about how you can understand yourself better when you learn about other people. They really did that—even with the fantasy aspect of it worked in.
—A third-grade teacher who had never made videos before

We show [our video] every year, in fact the other teachers from the team also want to borrow it and they show it before they go on the Erie Canal, part of their Erie Canal lessons. Some things you need to read in the book and show pictures, but the fact that there were certain things that were caught on camera—that really helped them understand.
—A fourth-grade teacher who had never made videos before

When I saw them paired up on the screen for each country, kids that were together were not kids that often played together in here, and I just thought it was a real nice way of getting kids to socialize better. I mean, it wasn't even part of the unit in terms of the social studies. . . . Everybody was very accepting of each other.
—A special education teacher who had never made videos before

None of these teachers had videomaking experience, but all witnessed the intellectual energy that it added to their learning environments.

Minds and Pathways

Both teachers and students benefitted from our videomaking experiences. Teachers found what was not there before—a tool to open pathways to learning they were not previously aware of. Though they were admittedly unsure of the method, these teachers were happy with the results.

Bringing cameras into the classroom serves as a conduit for children's visual expression—figuratively unlocking their moviemaking minds, letting them into the light of their classrooms. Perhaps most interesting of all was the fact that these moviemaking minds did not need to be created from scratch. They were there all along.

Note

1. A term he coined based on Brian Street's (1984) "Autonomous Model" of literacy.

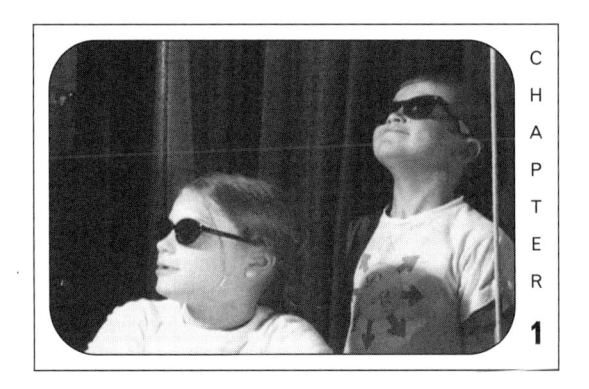

CHILDREN OF THE TV GENERATION

IN 1951, anthropologist Earnest Hooton[1] predicted television would "reduce mankind to complete illiteracy"—not to mention, it would also "ruin our eyes and our nervous systems." The radio and movies, he added, had already delivered a severe blow to literacy, and television would finish it off.

Fortunately, we live in an age where most people, such as second-grade teacher Diana Green, would recognize the sweeping exaggeration behind Hooton's claims.

> Ever since I was a kid people have been implying that television is evil. If you watch too much TV you'll turn your brain into bologna.... You know, I have some brain problems [laughing], but I don't think it's from TV.... I think kids can watch too much TV or too much of the wrong kind of program but I don't think TV or media is inherently bad. I don't buy that. I never have bought that.

This was a prevalent sentiment among the teachers and parents I worked with. Though they had no firsthand knowledge of negative effects of television, they were at least on the lookout for the negative effects that pundits had warned against.

It is such skepticism that has likely hampered TV's development as an educational tool over the past half century. Even when TV could be used to address learning problems, there was political pressure not to use it in the classroom.

For instance, Diana Green saw television as a comprehension tool in cases where students had reading problems. If a student had a reading (decoding) problem, it could stand in the way of her ability to comprehend, which could in turn affect her ability to write. Using television as a comprehension aid where reading was falling short could provide a vehicle for a student to write, while the reading problem was being addressed by other methods. Diana explained,

> Kids don't necessarily have to get their comprehension from print. They can go in and get it from video.... You can present materials through that medium instead of through printed material and do the same thing with it that you would do if you read the class a story or they read it to themselves.
> And I think up until now, the opinions have been ... parents don't want kids wasting their time watching video in school, and administrators don't want teachers using up time in school by putting on TV programs.
> So video's got this bad reputation, but it's there and it's not going away and this whole idea of visual literacy is educationally not addressed ... and the fact that it's not addressed is a problem because it can be an asset. So it's something that I think we need to all explore, because teachers I don't think are ever trained in this.

In this sense, comprehension—in particular, the lack of comprehension—was the problem, not television. Television is a possible *solution* to the comprehension problem. So why is TV not used across the board to deal with reading and comprehension problems?

Hierarchy of Media

This is where Kathleen Tyner's term "hierarchy of media" surfaces. She raised the issue as it related to an *Atlantic Monthly* study of standardized testing:

> It is true that most modern schools center around print literacy, but do constructivist practices and theories of multiple intelligences call into question whether or not print should be prioritized as better than electronic media in the pursuit of learning? It appears that the study assumes that print literacy is better than other forms of knowing and constructing meaning. I am curious to know if the researchers gave consideration to an assumed "hierarchy of media" as they designed and conducted the study. (1995)

What this underlines is the fact that print media are privileged in comparison to moving-image media in education. It raises the question that if we consider almost any reasonable form of print reading to be educationally beneficial, why do we stop short of the same reasoning when it comes to "television reading"?

Diana Green explained this hierarchy as both an administrative view and a parental view, and it made her and many of her colleagues nervous about using video:

> I think that sometimes video has a bad reputation.... Parents assume that teachers are using videos to baby-sit children and my feeling on that is that if parents are really upset about that, then that's how they use videos, and that's how they perceive videos at home....
>
> We use a lot of different things to get to children. We use videos, we use stories, we use field trips, we use teachers going in there and acting like crazy fools. But it makes a point and kids remember. That's what we're after. So yeah, I might put on *Aladdin* . . . but if I put on *Aladdin* today, it's because we're studying something that *Aladdin* illustrates....
>
> Then you have this parent saying, "My kid had two hours wasted on video and they've seen that video fifteen times and I don't want my kid coming to school watching videos."
>
> We've never had a parent come in and say, "Why did you read that book to my child? I read that book to my child at home."

In the minds of many parents and even teachers, books are seen as higher in educational value than videos. This seems ironic given the majority of parents and teachers today are products of the TV generation.

Children of the TV Generation

"TV generation" is a term normally used to describe the first generation born and raised with television—a group of people who essentially have never known a world without television in it. The term itself indicates the profound influence the introduction of television to society had, given its near-immediate designation as a generational icon. From the start, it seemed that people really believed the world would be drastically different with television in it—likely worse.

I remember grand theories floating around as I was growing up that predicted how the TV generation would be different from others—mostly along the lines of Mr. Hooton. It's actually not uncommon for such exaggerations to materialize when new technologies and trends are introduced.

In his book *Reading Audiences: Young People and the Media*, David Buckingham suggests such fearful branding of new ways of doing things is not unique to television. It is a defense mechanism to protect children and their "tender minds" from things adults do not sufficiently understand. Buckingham saw a parallel between parents' wariness of television and an ancient Greek philosopher's warnings about the ill-effects of poetry, which at the time was an unknown and highly controversial form of expression.

For Plato, and for many of our contemporary critics, these are essentially *educational* issues, which are to do with moral training and with building "good character." Yet their response is not to encourage young people to develop their own critical perspectives: on the contrary, the solution is seen to lie in greater control, and indeed censorship, by parents and by the State. Ultimately, what underlies these arguments—quite explicitly in Plato's case—is a concern to preserve the existing social order from threats which might seek to undermine it. (1993b)

Carolyn Marvin takes this a step further, suggesting that emerging technologies provide social opportunities to "negotiate power, authority, representation, and knowledge":

Media are not fixed natural objects; they have no natural edges. They are constructed complexes of habits, beliefs and procedures embedded in elaborate cultural codes of communication. The history of media is never more or less than the history of their uses, which always lead us away from them to the social practices and conflicts they illuminate. (1988)

One such conflict occurs with education. From the start, many educators saw television as a threat to established practices—a potential replacement for the face-to-face teacher. After all, television could allow one engaging teacher to reach far more people over short periods of time. Teachers I have worked with often expressed concern that they could not compete with television's popularity and flashiness, which they saw as a threat to the integrity of education. Education to them was *not* entertainment, but they worried TV's success might drain its intellectual integrity.

Now that *children* of the TV generation are in school, it seems we should have witnessed the negative effects, if any, television has had on its first generation. Television certainly hasn't destroyed literacy, as Hooton had predicted. It hasn't eliminated the need for flesh-and-blood teachers. It hasn't put bookstores out of business. Why, then, are some people still wary about the idea of using TV to teach their children? The simplest explanation is that the reasonably unfounded fear of television has been passed from the parents of the TV generation to the TV generation itself. The TV generation is afraid of its namesake.

The problem with this is that if the children of the TV generation continue this trend of fear, they will be missing out on the, up until now, untapped educational potential of television.

An entire academic school of thought—media effects—was in large part built around the fear of television and the potential effects it might have on society. Researchers seemed to sit at bay waiting for earth-shattering effects from physical to psychological to social. The *fear* of negative effects could in itself be a cause of what television has become, perhaps even more than the intrinsic qualities of the technology itself. Certainly television is a technology in a definitive sense, but technologies are also defined in their implementation, which is almost entirely a social factor.

The age of industrialization, which researchers often identified as a cause for profound social changes in the early twentieth century, was the focus of a strikingly similar wariness. Using the example of industrialization and its widely assumed influence upon social structures, noted social scientist Herbert Blumer explained the social tension generated by the introduction of technology. His central premise was that industrialization itself was the cause of nothing. Industrialization instead was the center of acting social forces that reflected the illusion of the social changes it was subsequently credited with. Blumer expanded on this notion:

> Precisely because no social happening is predetermined by industrialization but depends instead on the ways in which people mobilize themselves to respond to industrialization, social policies may play an effective role in shaping the social happenings, whatever it might be or in whatever area it might lie. (1990)

As television has been blanketed with no less credit for social change in the postindustrialization age, a comparison seems in order. This involves considering television as a largely neutral object of complex social constructions. Instead of monitoring social factors such as industrialization and new technology from a cause-effect stance, Blumer suggested the most appropriate vantage point for understanding such social phenomena is its "situations of contact" with social groups:

> Industrialization introduces situations [as does television technology] that make demands and set opportunities for new activities, new social relations, and new functioning arrangements. Whether industrialization induces social change depends on the nature of these demands and opportunities. The way in which people respond to the demands and opportunities sets the initial kinds of social changes that come into being. Neither the demands and opportunities that are introduced nor the definitions that lead people to respond to them in given ways can be deduced from the industrializing [or television] process. . . . Instead one has to see how the larger social process plays into the situations introduced by the industrialization process. (1990)

Television is a technology entwined in the complex landscape of social action. The point here is not to cast aside human defense mechanisms or discount the significance of the power struggle, but to see the dynamics around television and education from a more complicated position than simple causes and effects.

The established practices surrounding education have no doubt been challenged by the introduction of new media technologies such as television. The question of the role of media in education should be addressed by its educational effectiveness rather than by its threat level to established educational structures and tactics.

As Marvin advises, "New practices do not so much flow directly from technologies that inspire them as they are improvised out of old practices that no longer work in new settings" (1988). The problem shifts from criticizing the new practices to identifying the old ways that no longer work in new settings.

Guilt Complex

The "baby-sitting" quality of television Diana Green referred to earlier has also contributed to a guilt complex of sorts around the use of TV in classrooms, prompting teachers to ask, "Am I taking the easy way out by using television in a lesson plan?" I experienced firsthand how this guilt complex can stifle educational opportunities and innovation.

When I began formulating my research agenda in the early 1990s, I identified with an early media literacy school of thought. This included an assumption that there were dangers inherent in television content that children needed to be protected from. I set out to inoculate them from these dangers—dangers I had personally never experienced but others claimed existed. My strategy was to make them media literate by revealing the invisible practices, motives, and secrets behind media messages. If I could reveal this "underbelly" of television and teach children the proper way of watching it, they would be immunized from its dangerous effects.

Why did I build an agenda around an assumption that I deep down didn't believe in? The simplest explanation is *guilt*. Since I had built a career around making television, it only made sense as an ethically responsible academic to assume responsibility for the by-products of my career, which might be called "social pollution." Based on this vague and conditioned guilt, I was assuming that what I once did for a living had generated some level of social pollution. I had come to grips with this "reality," collected my "pollution credits" (since what I did was still needed by society), and now was going out to teach children how to best cope with their TV-infested worlds. The ultimate objective was to minimize the negative effects on them, if possible.

Origins

The idea for my first media literacy lesson came to me one day when my oldest son, Vaughn (five years old at the time), and I were playing around with a bendable desk lamp. I took out the video camera and showed him how to make the lamp appear to come to life with aid of stop-action videography. This involved recording a second or two of the lamp, stopping the camera while we adjusted the lamp to a new position, turning the camera back on, and recording another frame. We kept repeating this until we had enough footage to construct a fluid sequence of the lamp moving. I asked him, "Wouldn't this be fun to do in your class as a show-and-tell thing?"

He was all for the idea of me visiting him at school, and that was what I was really hoping for. As a new and still-learning parent, the idea of sending him out into the cold, cruel world of the schoolyard was worrying me. I liked the idea of being with him any way I could.

After several hallway conversations with his Mrs. Brighton, we came up with a day, time, and activity. Even though she had no idea how the children would respond to a media literacy lesson, she thought everyone would be better for the experience of trying it out.

Thinking back to my first session with the kids, it's difficult *not* to laugh. I conducted it almost like a college lecture. I made a point about how video worked, then showed them an example from a popular movie. They were a very well-behaved class, for kindergartners.

I was operating within a traditional media literacy paradigm, my objective being to arm the children with a critical awareness of media. Such a perspective should encourage them to question the media works they so often consume, ultimately making them healthier consumers of television messages. I translated my objectives into activities I thought they would understand and relate to.

I began by explaining that cartoons—and all films, for that matter—were actually groups of still images moving rapidly in front of a projector lens. We looked at a cartoon and examined some of its picture frames to make the point that cartoons were actually visual illusions.

But as my presentation went on they squirmed, their attention wandered, and overall they became increasingly indifferent to my lesson plan. The kids were far more interested in the cartoon's action and characters than whatever mystery behind their making I was trying to unveil. They acted almost as if I was interfering with their TV time and with their simple enjoyment of the imagery.

It seemed to me at the time that either I was incapable of reaching this particular age group, or the kids were not old enough for the concepts I was putting in front of them.

But when we started filming our own little "cartoon"—a stop-action "kid train"—their restlessness was transformed into positive energy. They couldn't get enough of the experience.

We set up a video camera on a tripod on one side of the room, and had the kids sit down on the other side of the room, one behind the other, arms wrapped around each other's waists. Once they were reasonably in order, I pushed the record button of the camcorder, recorded a couple of seconds of the "train" of kids, and then paused the camera.

At this point, the teachers had all the kids slide forward several inches and freeze for another video picture. Since the camera had not moved at all, and the kids had, our recording was creating the illusion that the "kid train" was magically moving around the room. We kept doing this until we had moved the train completely around the room and out of the camera view. This took a lot of coordination, tedious rearrangement, and raw patience to complete the exercise, but the kids loved it. They were far more responsive to the video production activities than to the lecture about how cartoons were made.

When we sat down to watch what we had done, the kids were thoroughly amazed with their visual accomplishments, particularly in seeing themselves on the screen. They also delighted in the cartoon illusion they had created. When I took a moment to try and bring them back to the media literacy point behind the exercise—the theory of persistence of vision and how everything on TV is not as it seems—I seemed to lose them again. All they could say was, "Can we watch it again?" Our exercise together was just a simple amusement to them that they didn't want to take seriously.

This little media literacy project was off to a slower start than I had hoped for, as these five-year-olds were eluding my "instructional snare."

Moving On Up

I didn't give up, though. I continued working with the kids as they moved to first grade, all the time searching for more effective ways of reaching them. I proceeded in largely the same fashion: explaining a media literacy concept, demonstrating it in a video exercise, and reemphasizing the concept after the project was completed. Our moviemaking was turning out to be a memorable and meaningful experience, but more as an action-based recreational activity than an intellectual one.

"The Magic Height Machine" was a good example of this. The point behind this exercise was to demonstrate the effect of lens angles in portraying a subject's size and power. The smallest child in the class was portrayed as the tallest, and tallest as the smallest. I talked briefly about the concept to them, then demonstrated it by wiring the output of my video camera to a TV monitor they could see. The point of the exercise was to demonstrate how the height of the lens with respect to a subject's eye level could affect the way that we perceived them as people. We would perceive people above the lens level as tall or important, people at the lens height as approachable or equal to us, and people below lens level as vulnerable or less powerful in relation to ourselves.

The kids could not contain their excitement. But it was the simple joy of seeing themselves on a TV monitor that thrilled them, not the point I was trying to make about lens angle portrayals. Their sheer delight seemed to be smothering the point behind my lesson plan. Nevertheless, I pushed on because even though I was not achieving expected objectives, the unexpectedly high energy around their participation made it difficult to write off the experience as a failure. Deep down, I felt guilty about all the fun the children were having with our television activities. The production activities were all working very, very well in terms of their ability to effectively engage the children. But the discussions about media literacy concepts seemed to have little impact. I really had to wonder if kids were simply incapable of being serious about television.

The most prevalent criticism I had heard over the years from teachers had to do with their fear of TV's entertainment factor. They worried they could not compete with television's flashiness and quick cuts in their educational tactics, and they felt a genuine pressure to compete with it. In addition to feeling threatened by it, they believed reducing lesson plans to flashy presentations cheapened the integrity and depth of education and contributed to a spoon-fed laziness—the constant need to be entertained. And these teachers worried that since they could not compete with television, it would slowly suck the life from the role of the traditional teacher, brainwashing children into complacency. I began to wonder if I was seeing evidence of such complacency.

With the second grade, I tried more structured media exercises. For Halloween we created a game called "Trick or Treat." The point behind it was to get the young participants to view entertainment programs more critically. We played a video recording of a program they were extremely fond of and talked about critical aspects behind its making—for instance, How are plots made up? How are scenes shot and edited? Why are commercials intertwined in programs? How are special effects achieved?

I then laid out the procedures of the game. If the kids saw something that was made to look real but was obviously not real, they would call out, "Trick!" If the kids saw any mistake made by the producers of the program—mainly, continuity errors—they would cry out, "Treat!" Treats were worth more than tricks because, even though they could be found in just about every program, they were less plentiful and more difficult to find than tricks.

The teacher and I were amazed when the children caught on to the game immediately. In fact, we couldn't watch the program more than ten seconds without someone stopping it with a critical observation. We had easily demonstrated that children were capable of critical viewing, which seemed like a major milestone for my work with them. It was actually *easy* for them to critically examine media and understand critical concepts I had thought might be over their heads. Perhaps this had something to do with their rising maturity from kindergarten to second grade. I wasn't sure why this game of critical observation was working, but I felt I might have finally broken through to them with my media literacy objective.

However, a few months later when I interviewed two of the most involved students in that class, I realized they had completely forgotten how to watch a TV program critically. They even forgot what a trick and a treat were! When I talked to their parents and siblings, they hadn't noticed any changes in the way the kids watched television. There was something troubling about these kinds of findings. It was as if I was not getting through to them at all. They were, it appeared, indifferent to my media literacy lessons! I had hoped that my work with them would activate a lasting critical sensitivity to media that would help them become healthier consumers of TV messages. This would have been teaching them the "correct" way to watch television. However well intentioned, my plan was not working.

It was at about this time in my research that I made a dramatic discovery—something that made everything that had happened in my media literacy experiences make sense. In the very simplest of terms, I realized that I wasn't listening to the children. They were listening to me, but I was not listening to them. And the ironic thing is that this discovery was triggered by television.

Enter TV Land

A flyer in my son's backpack announced that a major television network show was coming to our very own elementary school to talk to young children about their views on television.

I considered involving my own children. After all, the project was endorsed by the district's superintendent, and what harm could there be in letting kids talk about television? Then again, this was a national news magazine, a genre of TV programming not exactly recognized as a forum for harmless conversation. The news magazine shows I knew of tended to feed on controversy, drama, scandal, and ugly secrets. My inner debate continued beyond the participation deadline, so my children did not end up in the show's pool of five- to seven-year-old participants. When I watched the program some months later, I was glad that they hadn't.

The show was taped in an experimental lab at the nearby university. The children were involved in what appeared to be an "experiment" concerning their attitudes not just on television, as the flyer had indicated, but on *sex* as it related to television.

They were separated by gender and then placed in rooms equipped with hidden cameras and microphones. Cameras rolled while they interacted with one another and answered questions from an adult supervisor off-camera. There were some sexually suggestive media materials placed in their settings, including several fashion magazines and a supermodel video. They had no problem noticing and talking about all of the materials.

A group of three adults—the program host, a psychologist, and the head of a child advocacy group—spent most of the segment interpreting the young participants' behaviors and drawing broad generalizations about children, popular media, and sex. Together they continued analyzing the children's laboratory behavior, suggesting parents should work harder to control children's exposure to sexually suggestive media because young children were not at an appropriate age to deal with it.

When the program was finished, I felt angry and deceived—not because of its conclusion, which in some ways borders on common sense, but rather the way in which the conclusions were reached.

First, the show's producers had misled the parents of the children participating. If they were doing a story about children's perceptions of sex, they should have clearly indicated that during the recruitment stage. The school's principal explained afterward that had parents known the show's actual premise, most would not have allowed their children to participate.

Second, the show's supposedly scientific procedures were even more misleading than their recruitment tactics. It was inappropriate to draw sweeping generalizations based on the brand of sloppy science they were practicing. They seemed more interested in uncovering some sort of national epidemic— "If you scare them, they will watch"[2]—than in examining and better understanding a social phenomenon.

The premise of their story was that something new was happening among young children, something that challenged the notion that sex was over children's heads.

No social scientist I know would ever call this an experiment; it was more an entrapment than an experiment. Among the many holes in the process was the fact that the program staff ignored the impact of peer pressure and the unnatural environment of the university laboratory. It was irresponsible for the show staff to hype these findings. On top of it all, since it was my kids' school, I felt the sting on a personal level.

What did this whole news magazine experience have to do with my media literacy work with the kids? Though I didn't realize it at the time, the personal turmoil the program was putting me through also shook the foundations of my own work with the kids. Whether or not I bought the program's "sex exposé," witnessing the experience was revealing things about young children that would affect my interaction with them from this point forward.

Because I was watching someone else doing it, I slowly began to realize what I had been doing wrong with the kids those first few years. In our separate projects, the news magazine production team and I were both treating the children as naïve and powerless, and assuming ourselves as some sort of super-enlightened media experts. Neither of us was acknowledging the inherent intelligence, imagination, creativity, and sophistication of the children we were working with. We were treating them like sponges, as if all they did was indiscriminately soak up everything around them. Sponges don't move, don't think, don't play, don't imagine. Children do. Sponges aren't creative and unpredictable. Children are. Sponges won't resist their environments. Children will. Sponges won't talk back. Children do. I found myself asking, "Why do adults, myself included, tend to ignore these indisputably simple facts?"

The news magazine show had surmised kids were soaking up much more than most parents might think them capable of. They questioned kids' ability to handle information not meant for them. They did not consider what the information meant to the children, but rather how it threatened their own views on how much, or how little, children should be exposed to the subject of sex.

Perhaps knowing the kids on a deeper level enabled me to look beyond the superficial observations of the "experts." This was evident in the case of one girl—a girl who had been in one of the classes I worked with—who was asked to comment on a somewhat racy scene in a PG-13 movie.

INTERVIEWER
Does it bother you when you see that?

GIRL
Yeah, because if little kids watch it, like my little brother, he thinks it's okay.

The program staff focused on the fact that the girl, as well as most of the rest of the young participants, knew what was going on in the scene, knew that the sexual content was not flying over her head. What they failed to acknowledge was the girl's sophisticated reaction to the topic on a critical level, questioning its appropriateness for unaware minds. She was picking up on more sophisticated moral codes embedded in the issue of sex.

Interviewers would ask a question and children would answer it. The host would express some degree of shock over the fact that they had a response, but never really examined the response beyond its surface. Bits and pieces of children's observations were used to draw alarming conclusions like:

HOST
You may have heard that children are starting puberty at younger and younger ages. For some girls menstruation is beginning as early as age nine. Why? Possibly better nutrition, but does anyone doubt there may be something else at work here that is not just what children eat, but what they see?

Responses of the children were presented as some sort of embedded obsession with sex that the show's staff had managed to release from its hiding spot. The only thing the show staff had demonstrated to me was that they could successfully engage the imaginations of young children in a topic they knew very little about. The children, on the other hand, proved they could infer much about the topic based on extremely limited evidence and instruction.

The more I watched the program, the easier it was to see the children's sophisticated response to the topic and the program staff's ineptitude in dealing with them. They were trying to force children's inquisitive explanations and perceptions into a "Puritanical bottle."

Recognizing a Child's Perspective

In our own ways, the show's producers and I were inviting children to participate in essentially "adult discourse" and concerned by their "unadult" responses. We were too involved in our own agendas to see the interactions from a child's point of view.

Granted, our agendas were driven by notions of protecting children, but we were discounting the involvement of children in making our pronouncements. We were protecting them because in our minds they were defenseless, when in reality they were showing signs of awareness and strength.

I recalled images of the children restlessly squirming as I tried to deconstruct some sort of media concept to them. The problem wasn't just them understanding me, it was also me understanding them. I was approaching them as if they were blank canvases, with no existing position or understanding of television, just like the producers were treating them on the subject of sex.

As far as the magazine show's staff was concerned, television was doing too good of a job teaching kids about the "wrong topic." If kids could effectively figure out the dynamics of procreation from film and TV accounts, why couldn't they pick up other complex ideas, such as the ones I was trying to teach them?

I realized my approach with the children was built upon a complicated sense of guilt—call it the *TV generation complex.* This was partly related to my profession and whatever damage I felt I might have inflicted in my work, and partly as a concerned parent, not wanting to make any mistakes when it came to innocent, defenseless children. I was acting afraid of television even though I didn't believe it should be feared. I would need to shed this *complex* if I was to really get through to the kids.

What I really needed to do was to learn to understand their "language" and ways. If I was going to get through to the kids, I needed to approach the lesson plan as much or more from their viewpoint as from my own.

Back to Kindergarten

It was partly out of a sense of paternal duty that I returned to kindergarten to make videos with my second son's class. Aside from my growing interest in connecting media making with five-year-olds, the basic experience of physically sharing school experiences with my children felt right as a parent and as a member of the community. I was sending a sign to my children that school was important to me.

Not only had I come to know how exhausting a task it was to teach, or even supervise, large groups of young children, but I was also getting to know the children and feeling the pain of social inequity among them. An occasional one or two seemed further along intellectually than my own children, which was humbling, but more were at a desperate disadvantage in comparison. In short, whether or not I achieved my media literacy objectives, there was much to gain in the experience by simply being in there, and the possible benefits extended to and beyond my own children.

A Child's Approach

As a teacher myself, I knew that the maxim, "If at first you don't succeed..." applied to faulty lesson plans. Based on my kindergarten through second-grade lessons with my first son's classes, I decided I needed a fresh approach—a child's approach.

My mission remained the same: to meaningfully engage young children with media concepts—but my method would be different. This time I would try to follow the children as much as to lead, and see where they would take me.

This shift in strategy illustrates one of several debates[3] in the media literacy community, mainly one's response to the question, "Does media literacy protect kids?" My original response, and the news magazine staff's as well, was "yes." I had been schooled under the media effects paradigm, amidst a generation of great skepticism about television.

In my strategy shift, I would proceed from the assumption that children didn't necessarily require the inoculation of media literacy to be healthy consumers of media. I was being drawn to consider the other side of the debate represented by the views of scholars like David Buckingham. Buckingham saw the focus on media's problematic features as neglectful of the genuine pleasures children receive from it. In his view, adult superiority and cynicism over media should be weighed against some level of respect and appreciation for what media means to kids. Instead of teaching children the preferred, adult way to experience media, the approach would be first to find out what media means to kids and then how they make sense of it. This promotes an exchange of perspectives, rather than an enforcement of an adult view upon a child's existing views.

Children needed to be enlightened with a critical awareness of television to be healthy consumers of it, or so I assumed. Though I possessed such knowledge, I now had to wonder whether they already knew something about what I was trying to teach them. This could have been acquired by experience or even by some intangible and sophisticated inner sense. There were small indications that they did, for instance, when they demonstrated they could quickly catch on to the "Trick or Treat" game. At the very least, I had to rethink my missionary approach to saving the souls of young media consumers, and more carefully consider what media awareness they did possess.

Approaching media literacy objectives from more of a media appreciation and exploration perspective would provide children more space to operate. What drives activities of such an approach is not some media literacy credo—for instance, critical illumination of media practices and their effects on society—but rather how media fits into the expressive desires of children and their school experiences. This involves activating television literacy as a tool for lifelong learning, not unlike reading, writing, and arithmetic.

Everything I Really Needed to Know

Children have the capacity to connect to our worlds. The question is, should we regulate and possibly impede that capacity or cultivate it? Given the prevalence of media in the lives of children, it seems to me we have no choice but to cultivate it.

Recently I talked with Mrs. Brighton about the lesson I had learned about kids since my first media literacy lessons in her kindergarten class. I asked, "Why didn't you tell me I was so far off? I really underestimated the kids and their capabilities."

She smiled and said, "That's the way I learned. I think it's just one of those things you have to learn by trial and error. Besides, it was always fun and the movies are precious."

Certainly in my case, Robert Fulgham[4] was right when he said, "Everything I really needed to know I learned in kindergarten." I just didn't realize it at the time. It took me more than two years, three grades, and a news magazine exposé to learn what kindergarten teachers already must know—that there is a considerable intellectual capacity in young children that their everyday demeanor belies. Television can not only be a tool to stir that capacity but also to tap into their perspectives and perceptions that get lost when we try to fit them into adult worlds.

Notes

1. Quoted in the *International Herald Tribune*, May 2, 2001.

2. Jensen, Osborne, Pogrebin, and Rose (1998) coined this phrase to describe a tendency of newsmagazines "to overplay danger or heart-wrenching footage, the kind of viewer-grabbing hype that can obscure whatever caveats may be offered."

3. Drawn from Renee Hobbs's (1998) wider discussion, "The Seven Great Debates in the Media Literacy Movement," *Journal of Communication* 48, no. 1: 16–32.

4. Author of the book *All I Really Need to Know I Learned in Kindergarten* (New York: Ballantine Books, 2003).

CHAPTER 2

READING, WRITING, AND TELEVISION

A new medium is never an addition to an old one, nor does it leave the old one in peace. It never ceases to oppress the older media until it finds new shapes and positions for them.

—Marshall McLuhan, 1964

"**WE WANT** to use video to get young readers excited about reading!" said the voice on the other end of the phone.

Most of the calls I get about video production have to do with preconceived ideas for shows or promotional messages or films, and the callers want technical assistance to bring their ideas to fruition. This one was different—excitingly different—because video is rarely associated with the coveted practice of reading. The two are usually thought of in distinctly separate circles, especially by educators, and this was a call from the superintendent's office of a large city school district in Pennsylvania.

The Romantic Ideal of Reading

> Reading is to the mind, what exercise is to the body. As by the one, health is preserved, strengthened, and invigorated: by the other, virtue (which is the health of the mind) is kept alive, cherished, and confirmed.
>
> —Joseph Addison, *Tatler*, no. 147, 1710

If there's one thing school teachers seem to agree on, it's the value of reading. It seems to be a universal law of learning. "You can't read enough books! Reading will expand your world!" Every summer, the marquee in front of my son's school reads "Summer is for Reading." In bold letters on every page of the English Language Arts K–8 curriculum book, formal statements like the following can be found: "At all grade levels, students are expected to read (or have read to them) twenty-five books each year."

It's one of the very few things in life that seems to be exempt from the law of "everything in moderation." What is so essentially good about reading that teachers feel their students can never do enough of it?

Reading is to learning as breathing is to living. Reading is a tool for children to accumulate and build knowledge. The New York State Learning Standards[1] situate reading (with writing, listening, and speaking) as tools

- for information and understanding
- for literary response and expression
- for critical analysis and evaluation
- for social interaction

Although the curriculum includes a sampling of appropriate books for each grade level, there is no mention of appropriate television programs, videos, or films. There is also no mention of video cameras and how they can be used for writing and expression. The suggestion here is that moving-image media are not seen as legitimate learning resources.

This is an example of how K–12 curricula tend to cling to traditional print texts as reading resources, when alternative reading sources could expand students' exposure to ideas and contribute to a wider knowledge base. Quite simply, moving-image media texts are not officially recognized as reading texts.

The problem with this is that educators are submerging a vast and compelling learning resource when they don't acknowledge television as a resource for learning. This is why I was surprised when I got a call from someone in education who thought TV should be used to get kids excited about reading.

The Reading Project

The superintendent of this city school district was widely known as a reformer. David Harrison was not afraid of questioning the status quo or trying new techniques to invigorate learning.

The district had just received a grant to be used toward improvements in literacy initiatives. Its major thrust was to find a way to help children to see reading more as a lifelong pleasure than as a formal academic act only applied in classrooms. There were three objectives behind the grant:

1. To build a stronger relationship between the newspaper and young readers
2. To assist the *Books for Kids* project in expanding their reach to those children most in need of their services
3. To widen the reading universes of young readers by pointing them to reading experiences beyond the mainstream

Dr. Harrison saw this as an opportunity to make connections to novel projects underway throughout the district. He had already put together a program to revitalize several high schools with themes of specialization, the newest of which was a school focused on media. He believed television and the Internet could and should play a role in the foundations of education.

The media-centered high school was Dr. Harrison's pride and joy, one he hinged many hopes on. Lakeshore High was located in one of the poorest neighborhoods in the city and had been neglected for years. The new building was built within view of the old, almost in defiance of the challenging inner-city elements that surrounded it.

The new Lakeshore High contained several media features, including a state-of-the-art computer lab, a television studio, a radio lab, and video-editing stations. Dr. Harrison believed the reading project could be worked into the media mix, demonstrating that media could play a role in strengthening literacy in the district.

He and his assistant shared their vision of a television program that might be produced with the grant and asked what I thought of it.

> We could use high school kids as the producers of a show where they read selected books and sections of the newspaper to younger kids. Hopefully the experience would raise the curiosity of younger children and they would be inspired to read the book of the week and warm up to the newspaper.

TV 101

One of the first lessons students of television production learn is how limited a medium TV truly is, especially when it comes to traditional reading. First, television does not effectively or efficiently render printed text. TV's small screen size and reasonably low-level image quality do not handle text well. Letters must be very large (usually 48 points or higher) to be read and this limits the amount of words per line and lines per page that can be used to communicate print-based ideas.

Second, having people read stories to a television audience in the same way they might read them at a story hour in front of a class is generally ineffective in engaging any TV viewers, let alone the youngest and most restless of all viewers—children. For this reason, TV production students are drilled to *show* with TV rather than to *tell*.

These two factors demonstrate why television can never hope to replace books as a delivery medium of printed information. Like any other medium, TV has strengths and weaknesses. TV's strength is in action and movement, not letters and oratory. Therefore, to utilize television in reading literacy, users must understand what television can and cannot effectively contribute in a lesson plan.

Even though TV has not replaced the medium of books as many educators once feared, it has added a different dimension to the experience of reading in the late twentieth and early twenty-first centuries.

TV as "Reading"

read•ing, *n*
to receive or take in the sense of (as letters or symbols) esp. by sight or touch

Consider this. If reading is good for people, then so is television—because television content, like book content, is a text. The difference is in the decoding process: letters and words versus pictures and sounds. Both texts are used to form meaning and compile information into shared experience, knowledge, and understanding.

This seems like a silly idea to most people, partly because we have all been conditioned to be wary of TV content, especially in educational circles. But there is also something to be said about the ease in the TV reading process as compared to the printed-text reading process. Children learn how to read television much earlier in life than they learn printed language.

Perhaps deep down we feel guilty that something so easy shouldn't be *good* for us. Learning to read books was difficult and time-consuming work, and we were praised for reading books. We were not praised for watching television, even though much of it required skill and imagination to interpret.

Media scholar Marshall McLuhan (1964) distinguished between the decoding methods of television and more traditional media by their respective temperatures. *Hot* media such as books, radio messages, or films were in his words "high definition" or loaded with sensory-specific data with little interpretive wiggle room. Hot media therefore required low levels of audience participation since messages within such media had reasonably fixed and intrinsic meaning.

Television and telephone (and most certainly, if McLuhan were alive to update the list, the Internet) were examples of a different media temperature—*cool*. Cool media were comparatively low in definition with wide spaces for interpretive audience involvement. Since messages in such media required high levels of interpretation, audience participation level was seen as high, in McLuhan's view.

A typical commercial, for instance, requires a viewer to build meaning through the juxtaposition of images and sounds—meaning that is not inherently *contained within* the message, rather *inferred* from it. If we see a child playing with a toy in a commercial, and the child looks at something, the shot following the look is inferred to be what the child is looking at. A viewer is cued to take on the point of view of the child and see the object from that child's perspective. On a more macro level, if a short and entertaining story is juxtaposed with a graphic of a product name (for instance, Federal Express), a viewer will easily associate the meaning and outcome of the story with his or her impressions of the product name. These are just simple examples of visual montage, but they demonstrate the need for reading skill in watching television. Unfortunately, this reading process is given little intellectual weight in educational circles compared to the act of assembling words into imaginary images in a book reader's mind.

The experience of "reading" television also takes place in a wider space; as opposed to the distance between eyes and a book, television viewing takes place between viewer and television, and there is more opportunity for interference in that space. People often watch television with other people, whereas book reading (outside of reading to others out loud) tends to be a more solitary act between book and reader. TV viewers often engage in simultaneous activities—knitting, talking on the phone, reading books, doing homework, conversing with friends and family—all while they are watching. Interestingly, such simultaneous activities don't necessarily mean viewers will have problems comprehending television content. In a study with five-year-olds, Anderson, Alwitt, Lorch, and Levin (1979) discovered that children who watched television with distractions (toys in their case) watched a *Sesame Street* episode for 47 percent of the time. A second group in a room with no toys watched the show 87 percent of the time. The big surprise was that both groups scored identically on memory and comprehension. This suggests that kids were capable of doing sophisticated multitasking.

The End of Reading as We Know It

The doom of the reading habit has been falsely prophesied ever since the invention of the pneumatic tire which spelled the end of the fireside reading circle by putting the whole family on bicycles.

—S. B. Neuman (1991, 375)
quoting R. D. Altick, *The English Common Reader* (1958)

Many teachers I have talked to believe that TV strips imagination from reading because it provides so much imagery for viewers. They felt that the ease of reading television made learners lazier and discouraged them from reading books.

S. B. Neuman cast considerable doubt on this myth in her book *Literacy in the Television Age*:

> What is truly intriguing is that television has never been shown to displace reading. Even during its novelty phase, studies by Himmelweit, Schramm, and others did not report changes in leisure reading. Indeed, these studies indicated that there was little reading time for television to displace in the first place. Time spent leisure reading has remained surprisingly stable over the years: There was not much reading before television, and there is not much reading today. (1991)

Ironically this implies "television reading" has become an "above and beyond" knowledge resource in an area that textual reading has not been able to penetrate.

What skeptics of TV don't recognize is that there is still imagination required in creating meaning out of moving images and sounds. Diana Green elaborated on this in the context of her second-grade reading curriculum:

> If kids can't visualize—if you read a paragraph about the blue sea, and the bright sun on the sandy beach and the kid can't draw that in a picture or at least explain to you what it should look like—we find that's also the kid who can't compose, who can't get comprehension. Visualization is really important.... We do know that kids who can't visualize, can't read.

Diana saw video's potential in cases where students had difficulty visualizing print-based stories. Researchers have uncovered similar links between reading strategies, visualization, and comprehension.[2]

Gavriel Salomon (1982) was able to demonstrate that comprehension, an ultimate objective of any reading experience, need not be considered media specific. He described three steps to comprehension:

1. Mental recoding of a message into a parallel mental representation (what Diana refers to as visualization)
2. "Chunking" or integrating these elements into meaningful units
3. Making elaborations on that material—drawing inferences, yielding new attributions or questions

He argued that as the stages of comprehension progressed, media-specific skills (e.g., print reading skills or television reading skills) mattered less. Since the main objective of early stages is to efficiently and effectively receive information, it stands to reason that the easier information is to collect, the more successful later stages are likely to be.

Since television literacy is normally acquired by children before print literacy, it can contribute to overall learning as a comprehension aid. Messaris saw such visual literacy as innate:

> Unlike conventions of written language or, for that matter, speech, pictorial conventions for the representation of objects and events are based on information-processing skills that a viewer can be assumed to possess even in the absence of any previous experience with pictures. (1994)

In this light, television can be seen not only as an aid to comprehension difficulties but also as a legitimate primary reading and learning resource.

The Ease of Reading

Certainly, *all* reading is not good for *all* people, and there are limits to how much time a healthy person should devote to the act of reading, as opposed to physical activities and social interaction. And so it is with television. But many, if not most, educators tend to scrutinize television "reading" as if it will hypnotize viewer's sensibilities and lure them into endless hours of mindless consumption. This is because most educators, unlike McLuhan, tend to downplay the intellectual depth of television "reading" because of the comparative ease of its "reading process."

Neuman borrowed information processing theory to sketch out a ground somewhere between McLuhan and skeptical educators. She surmised, "The information processing demands from television and reading, although they rely on different codes, are equivalent and serve a similar role in knowledge acquisition" (1991). This suggests that regardless of medium, reading is reading. Regardless of whether "reading" occurs with books, films, radio, the Internet, or television, it can be applied to the underlying goals of education: information and understanding, literary response and expression, critical analysis and evaluation, and social interaction.

The *ease* of reading television and other popular culture texts provides the opportunity for children to read more and ultimately achieve the objectives underlying the act of reading: to learn and understand more about the world.

Kathleen Tyner elaborated on this point in her book, *Literacy in a Digital World: Teaching and Learning in the Age of Information:*

> The use of popular culture texts in the diverse classroom offers a promising intervention strategy that explores the nature of discourse, bridges cultural understanding, and enables students to realize the full range of expression available to them. Student knowledge of discursive modes and their relationship to media through a combined production and analysis approach is

intended to lead to a deeper understanding of the more traditional, sanctioned texts that are valued in the cultural canon of the dominant culture. Student readings of such texts may be "against the grain," but will at least result from a reasoned approach to appreciation. (1998)

Just because something is easier for children to read doesn't mean it has less intrinsic information value for them. Such reading should not be seen as a replacement for traditional reading but rather as a complement to it.

If educators can accept the viability of television as a reading resource, the issue becomes one of finding the right function for TV as a learning tool. Neuman elaborates on this:

Interests in one medium tend to be reflected in the other. Rather than compete, there is spirited interplay between print and video activities that may spark children's interests and enhance literacy opportunities. Thus the notion that there is only one road to literacy is culturally derived. Television provides a wide variety of fare that when used appropriately, has the potential to complement and enliven literacy. The responsibility and the challenge of using television to expand children's learning and literacy, however, lies not in the technology, but in our hands. (1991)

In this sense, the use of television in classrooms is a matter of teaching, not technology.

Developing the Reading Project

I posed this very point—that the use of television in classrooms is a matter of teaching, not technology—to Dr. Harrison as we considered new directions to take his idea of using television simply to read books to elementary-aged children. It didn't take long to demonstrate to him why reading books on camera might not work. Based on my experiences with videomaking and kids, I suggested that we put the challenge to kids in his district. The schoolchildren I had worked with over the years had always thrived on creative challenges I posed to them.

Dr. Harrison liked the spirit of this idea and it made him think about his video tech program at another of the district's high schools. This program was run by Thom Samson, an independent TV producer turned teacher. He and his students were making some waves through the district with their production work. David regularly called on Thom to videotape projects and events and to air them on the district's cable channel. He was very impressed with Thom's initiative and results with the students.

Since David did not yet have a curriculum or video person in place at the new media-focused Lakeshore High, perhaps Thom would be a good person to lead it, and this reading project could be the pilot effort for a new program.

At this point, we met with Thom's two video-production classes, students from grades 9–12, and put the challenge to them: how might television excite elementary-aged kids about reading? As expected, they didn't hesitate, and immediately brainstormed ideas for shows they would have loved to watch when they were learning to read.

Each of Thom's classes came up with its own show and agreed to live with the sponsor's decision on which was best to produce. The first group proposed a fantasy/action series involving a group of superheroes called *The Read-A-Lot Gang*. These characters would be engaged in a series of good versus evil plots involving reading. Through their endeavors, viewers would see the value of reading and have the chance to get involved in the story by doing reading themselves. The more viewers read, the more power the superheroes would have.

The second group proposed a magazine-style program called *Imag-A-Book* that would highlight books of the week with student interviews, reenactments of book segments, and conversations with literary experts on reading resources near and far.

Both groups were extremely excited about their respective ideas. I was surprised how seriously they had taken the assignment and how involved they were in, essentially, *someone else's* idea. I had worried they might see this as some "official school project" with questionable relevance to their comparatively more imaginative worlds.

It was difficult to choose one of the ideas over the other. My immediate reaction was to side with the more "rational," grown-up magazine program. I worried that the *Read-A-Lot Gang* might stray over time from the objectives of the grant, if the kids got too wrapped up in the fantasy. The *Imag-A-Book* approach seemed more tailored for the traditional notions around reading, not to mention, it likely would appeal more to the sponsors who were involved in the newspaper and book publishing industries.

On the other hand, I remembered some of our best experiences over the years of classroom videos had been the fantasy works students had dreamed up.

Reading Popular Culture

The question here was, how can a fantasy movie connect to a word-based activity like reading? Building on the work of Clark and Salomon (1986) and David Olson (1977), Neuman suggested that alternative media portrayals not only convey knowledge but create pathways to new levels of knowledge, concluding "television, employing its unique symbol system, will yield a different set of skills in acquiring new knowledge" (1991).

If educators can recognize television as a legitimate reading resource, learning opportunities abound. First of all, one of the distinct aspects of television "reading" is its natural, social orientation. Television watching is naturally communal. Book reading is more solitary in comparison.

If students have book-reading problems, teachers are equipped to deal with them with everything from special testing to reading assistance. Since television is not seen as a legitimate reading resource, children are left to read TV on their own. But the reality is, children need the same kind of nurturing in reading television as they do in reading books. Teachers are a sensible resource for such help, and the classroom is a safe and healthy environment to complement the process.

First-grade teacher Sally Beechwood recognized the need for this when it came to her students and scary movies:

> Kids will come into class and say "I was scared last night, I went to see *Friday the 13th.*" I would think, "Why would you take your child out so late and why would you take them to this?" I don't know, I wouldn't do that with my own children. I think some of these movies need to be explained a little bit. I think kids are genuinely scared. They take a lot of it at face value, this is the way it is and it might happen to me.

This is an issue beyond censorship. Even if parents think there are no visible effects on their children, teachers are often in a position to see them. Although Sally preferred censorship, she was powerless to enforce it upon her student's parents. Instead, she provided a refuge of togetherness, sharing, nurturing for her students. She was there to help her students "read" the media that confused or scared them, but scary movies were not the only thing she needed to help them with:

> They are scared by what they hear happening on the news. "Mrs. Beechwood, did you hear that a mother drowned her two kids?" They pick up on that. The bombing—it's hard to reassure them. "Well, you know—that was there and we're here and your Mom and Dad will protect you." But sometimes you think you're lying to them, because it could happen here. It is scary and they pick right up on that bad stuff. It is hard, because I don't want to put my own morality into it. I try to be moral, but neutral. I don't want to offend any parents. That's real touchy.

Children are keenly perceptive of the world around them, but don't always share this with their parents. Perhaps they are "protecting" their parents by not sharing their anxieties with them. If parents participated more in their television experiences, children might be more likely to share their anxieties with them. The bottom line here is that children know the world is dangerous no matter how much parents may care to insulate them. When children's access to ideas and imagery that confuse them increases, teachers can be a resource in helping them come to make sense of such media. The key is to see this issue as a reading and learning opportunity.

Reading into Popular Culture

In addition to helping children read scary or confusing media messages, teachers have the opportunity to interact with kids as they make sense of popular culture works specifically designed for them. Diana Green saw this in her second-graders:

> Each crop of kids has their own things. I know when my kid was growing up it was *Star Wars*, which probably isn't much different than *Power Rangers*—you know, good and evil, lots of action.

Getting involved with children as they participate in their "crop-specific" experiences not only provides children opportunities for understanding and clarification but also could help the teacher understand the unique perspectives of his or her students, and better connect with them as a teacher.

But there are limits to the involvement level that teachers can expect with kids when it comes to the popular culture works they identify with. This is because children often use popular culture to distance themselves from controlling forces in their lives.

Kathleen Tyner advised a middle ground:

> If teachers archly criticize popular communication forms, they run the risk of alienating and insulting the very culture their students value. If they embrace it, they risk looking like ridiculous fuddy-duddies who are trying to appear up-to-date. The middle ground is to follow the students' inclination to explore popular-culture themes but to use popular culture to guide students to more sophisticated, investigative, cultural pursuits. In the course of walking this tightrope, it is not necessary for teachers to suppress personal distaste for popular culture artifacts, or to express glowing enthusiasm for every new pop culture riff that comes down the pike. (1994)

In the end, works of popular culture, including television, provide not only additional reading resources but also different reading contexts and approaches to learning that could open up new pathways to learning. Like music, in the words of composer Roger Sessions, "The most precise possible language ... it achieves a meaning which can be achieved in no other way" (Gross 1974). Visual media provide respectively different filters to illuminate knowledge.

Imag-A-Book versus Read-A-Lot Gang

The "middle-ground" approach would come in handy in making a decision on which of the two reading project proposals to accept. After talking with Thom, we agreed that it would be difficult to choose one over the other and still expect both classes to contribute. So we asked both groups to consider combining their shows into one. They were very happy with this decision and we immediately mapped out a plan.

The two video classes would join forces to produce six episodes of the series, *Imag-A-Book and the Read-A-Lot Gang*. One class would produce the Imag-A-Book segments of the show, which would serve as the structure of the entire program. There would be two hosts—a high school boy and girl—and together they would introduce the various segments of the program, including the featured superhero segment, the Read-A-Lot Gang.

The Read-A-Lot segment would be produced by Thom's other class. This would consist of four main characters with "superhero powers":

- Goldguy: A masked and caped superhero and book lover.
- Cool Guy: Fellow book lover and good-natured "Fonzie-like" friend of Goldguy with a heavy Brooklyn accent.
- Sirac: Evil villain plotting to rid the world of all forms of reading. He was a proud graduate of Evil Villain Training School.
- Poison: Sirac's loyal sidekick, from a family of "no-gooders." She believes she is incapable of getting beyond her "no-good" label. Her continual curiosity about books and reading creates conflict.

The gang had a host of support players who would be swept into the battle to preserve reading as we know it. Sirac and Poison were usually involved in some scheme to rid the world of books and Goldguy and Cool Guy were never too far from the action.

All the show elements were to be tied to particular literacy objectives determined by the sponsors and superintendent. These included the following:

Newspaper in Education Segment

This was a children's story of the week printed in the comic section of the local newspaper. The newspaper would supply the high school production team with a hard copy of this story two to three weeks in advance of publishing. This would allow them to produce related program segments in advance of delivery. Sections of the newspapers would be distributed to the district's elementary schools on a bimonthly basis.

Educational Objective: To help kids see the newspaper as a source of reading for them, not just for adults.

Creative Solution: The kids decided the best way to work this into the program was through the Read-A-Lot Gang. They built a feature into the program that tied viewer reading to the plot. The more viewers read, the more power Goldguy and Cool Guy would have to overcome Sirac's evil plan to rid the world of books. The production team designed and delivered gold boxes to each of the district's elementary school libraries, where elementary students could place their answers to questions about the stories in the newspaper. Children who left messages had a chance of having their names read on the air by Goldguy in the next episode.

Books for Kids Segment

The organization Books for Kids would supply one of its current offerings to a particular elementary class each week. Students would read their particular book (or book portions) together and perform some sort of reading comprehension activity (for instance, draw pictures of their favorite aspects of the story). A field crew from the video class would be sent to these classes to document reader perspectives on the story. The crew would also get a shot of the entire class to be used in the segment's opening.

Educational Objective: To make the Books for Kids organization more visible to city school children and to identify more children in need of free books.

Creative Solution: The kids decided to create a "magic book" segment to the show and reenact a section from it to pique viewer curiosity.

Book Beat

This portion of the program would call attention to exciting, out-of-the-mainstream reading opportunities for young readers. This would be an opportunity to widen viewers' reading universes and promote attention to diverse perspectives (cultural and ethnic voices), timeless classics, and special experiences highlighted by kids, teachers, literary experts, and visiting authors.

Educational Objective: To expand reading universes of young readers and promote lifelong reading and learning beyond school.

Creative Solution: Originally, the students planned to connect with a local book expert. When that fell through, they decided a better solution would be to visit elementary schools and talk and listen to kids about their reading. They also had a "reporter on the scene" to interview kids about the books they were reading.

Benefits for Viewers

- Increase awareness of reading opportunities
- Build cultural awareness

- Broaden literacy by linking television viewing to reading experiences
- Foster a love for reading

Benefits for Video Students

- Hands-on experience in television production
- Opportunities for student actors to strengthen oral communication skills
- Confidence, poise, and professionalism through accomplishment
- Strengthening critical thinking, by putting students in the position of teachers
- The opportunity to work as part of a team
- Teaching students how to receive constructive criticism and use it to improve their work

It wasn't long before the students had assembled their pilot episode and were ready to present it to the superintendent and sponsors. I was nervous for them, but very proud of their accomplishments. I had been hired as a consultant to the district to implement the reading project, but I was feeling more like a co-teacher with Thom than a consultant. The students were taking ownership of the project and blossoming in front of us, and not just in terms of their video coursework. They were maturing as writers, readers, listeners, and human beings. As they put together their scripts for the reading project, they were becoming better citizens of their community and teachers themselves. They were captivated by the prospect of creating a show that young children watched and learned from.

This was a quality I had recognized in my professional colleagues over the years, as well as in the students in my university classes. People in the business of television and film production tend to share a deep interest in impacting other people with their ideas and productions. They want to make others feel the way they have when they have watched a great film or television show—in essence, to give back the gift of storytelling that they have received over the years.

As I watched the students busily preparing their pilot presentation, I realized they had breached the barrier between TV reader and TV writer. This is an important barrier to understand when it comes to utilizing media production as a curricular tool.

The Problem with Read-Only TV

When comparing the textual properties of print and television, there is a troubling disparity in their reading and writing characteristics.

Both forms of communication are amply read, but only print is routinely written. Imagine what the world would be like if we only read print, and had no means to express our own ideas unless we were part of a professional class of writers. Unfortunately, this is the structure in place when it comes to visual media. We can read the industry texts of television and cinema, but most who read these texts lack the means to "write" their own. Thus, most new media are "read-only" in an educational sense. This is an area ripe for reform and research, as Buckingham explained,

> We need to know much more about how students acquire academic discourses, and what the consequences and limitations of this might be. Furthermore, any educational conception of "media literacy" will need to consider children's own media productions as well as their use and interpretation of existing media. It will need to look at children as "writers" of media, rather than just as "readers," and at the relationship between these two sets of practices. (1993a)

Printed discourse is introduced to children as a means of two-way, inclusive expression—to be read and/or written. Power derives from the ability to write in a medium. Readers of "one-way media" are effectively in the dark on the procedures and motives behind the messages they receive. This puts the form at odds with children and thrusts educators and media literacy advocates in a position of defense.

The obvious next step is to open one-way media up to an interactive flow—to expose children to media "writing" skills and empower them through participation in such media.

Discourse surrounding popular media (music, cinema, TV, magazine) should be approached in an inclusive manner, like print, that emphasizes participation, expression, appreciation, and critical examination rather than on-guard skepticism.

Writing for Reading

Writing not only demystifies a medium, but from Diana Green's standpoint, it helps make better readers. One of the most important experiences of her second-grade class was the book writing project. A large part of her students' year was spent assembling all the story components. Even though she recognized the inherent value and novelty of students writing their very own books, there were other educational motives behind the project. She explained,

> Children who are good readers or successful readers, or who get a lot of comprehension out of what they read, become good writers because they understand that when you read you have to have comprehension and the other side of the coin is that whatever you read, somebody had to write.
>
> So when they start to write, they understand this relationship that information has to be given in a logical, understandable way. Children who are good writers therefore become better readers because they know how

to look for information. It's like a double-sided coin and they both need to be there.

Diana realized a similar dynamic at work with the making of her second-graders' movies. The reading/writing interplay crossed media when the students were scripting the movie.

> Having the opportunity to write the [movie] story as a group and see it produced and come together made the kids' experience in writing their own individual book go a lot more smoothly and successfully.

Adapting the storytelling process to the collective experience of making a movie gave Diana a chance to emphasize important aspects of the writing process.

How Writing TV Can Change Reading TV

Diana also noticed that the TV writing experience had changed the way the kids and their parents and teachers thought about TV. Diana related,

> I always get the same reaction: people laugh, people get kind of emotional about it, and then they just say all these fabulous things about the movie. And then I ask, "What is special about this movie?" ... It made it OK for kids to watch TV. And I think in many instances, people give kids the idea that it's not OK to watch TV, or it's not OK to watch certain TV or a certain amount of TV. But nobody ever said anything about this being limited in any way, that you could only watch it once, or that you could only watch it for an hour. This was OK TV. There was nothing wrong with this, nobody had any stipulations on how it was to be used, and it was just wonderful TV.

This was because it was *theirs*—conceived and created by them. They were participating in the "conversation" of television—not just taking it in but also giving it back. As participants, they had power, they had reach, and their voice was being heard.

Reading, Writing, and Television

I couldn't help think the same thing as I watched the superintendent and sponsors congratulate Thom's video students on their excellent concept and production. They seemed as proud of the students as Thom and I were, and thoroughly amazed with their energy, enthusiasm, and creative and technical abilities.

Not only was it well produced, but there also was an extra-special quality about a show celebrating reading, and the notion that it was something written by older students for younger students. It lent a special authenticity to the content because it wasn't a message from adult teachers.

Behind the scenes, the video-production students were proud of themselves. They were having an effect on kids and adults, and they were doing a lot more than just playing around with cameras and microphones and editing machines. They were being challenged on all fronts: reading, writing, and television—and they were loving it.

Notes

1. From *Key Concept Guide for Parents: Grades Pre-K–8,* a publication of the Office of Curriculum and Instructional Services, Syracuse City School District.

2. See David Buckingham's (1993a) *Children Talking Television: The Making of Television Literacy* (p. 29) for useful synthesis.

CHAPTER 3

VOICE OF THE DEVELOPING MIND

I dreamed I stood in a studio
and watched two sculptors there.
The clay they used was a young child's mind,
and they fashioned it with care.
One was a teacher: the tools he used
were books and music and art.
One was a parent with a guiding hand,
and a gentle, loving heart.
Day after day the teacher toiled
with touch that was deft and sure,
while the parent labored by his side
and polished and smoothed it o'er.
And when at last their task was done,
they were proud of what they had wrought.
For the things they had molded into the child
could neither be sold nor bought
And each agreed he would have failed
if he had worked alone.
For behind the parent stood the school,
and behind the teacher, the home.

—Author Unknown

WHEN I FOUND the poem in my son's backpack, I was inspired. I liked the way it defied the schoolhouse myth that learning is a one-way path of illumination from teacher to student. It conjured up what I thought at the time to be a noble image of learning—teacher and parent working together to mold the "clay" of developing minds.

So I saved the poem, if for no other reason than to remind me my work wasn't complete when I arrived home from the office at the end of the day. It has served as a continual inspiration to get more involved with my kids and their teachers and classmates, especially when the effort required becomes inconvenient in the grind of day-to-day life.

Beyond my videomaking activities, I ventured into some more demanding roles, including guest story reader, helping hand for craft day, field-trip chaperone, and such. In these roles I came to understand and appreciate the exhausting challenge teachers face in reaching all their students. It had always appeared so simple, so romantic, from a safe distance outside of the classroom.

And the more I worked in the classroom, the more I realized the poem that helped inspire me to be there was missing something in the whole sculpting process—a vital ingredient. Beyond teachers and parents there was a third, sometimes invisible, but incredibly powerful force involved in the process—the unique and creative voice of the developing mind.

Kids and a Camera

The first time I realized this, I was in the middle of a moviemaking lesson with Mrs. Beechwood's first-grade class. I had plans to teach them a lesson on camera composition: how to frame a shot, how to use focus and angle for impact, how to work with light.

I connected my video camera to the TV monitor in the classroom so the children could see what I was taking pictures of. Without thinking, I pointed it at them and the moment they saw themselves my lesson plan was over. I had tapped into so much energy and enthusiasm that I could have spent the whole day simply showing them images of themselves. They were captivated by the experience.

They took the opportunity to "ham it up" while they were in the spotlight, making faces, laughing loudly, stepping in front of each other, and overall enjoying every moment of the attention they were getting. When Mrs. Beechwood and I managed to calm them down and get back to the discussion of composition, the children had trouble focusing on anything other than the attention the camera was giving them. They were starved for the opportunity to be seen, heard, and noticed. And this, as I discovered through the years, was not just a first-grade phenomenon—it worked at every grade level.

In the case of this exercise, the "clay" I was trying to mold was not responding to my sculpting. They had their own designs on this lesson plan.

Understanding the Developing Mind

> We do not know how a butterfly works. We do not know how an egg becomes first a caterpillar, then a chrysalis, and then a butterfly—only that it does. . . . At our present level of knowledge these developmental and behavioral processes may as well be magic, and therein lies the source of mystery, challenge, and excitement.
>
> —Kjell Sandved and M. Emsley, *Butterfly Magic*

There is certainly no less magic, mystery, challenge, or excitement involved in the human metamorphosis from infancy to adulthood. The term "developing mind" suggests an act in progress, framing the child as a human work in progress. This process is something adults/parents/teachers tend to be extremely guarded about, tirelessly defending our children's safe and prosperous development.

The teachers I worked with often described their young students as "sponges," given their reception to new knowledge was exceptionally keen. But as Sylwester (1995) cautions, although infants can easily master any human language, they are born proficient in none. Their brain development and education in general could therefore be seen as "a dynamic mix of nature and nurture."

On the side of "nature," Buckingham cautioned against

> a view of young people as a unitary or homogeneous social group, with specific psychological characteristics. Most obviously, this involves paying close attention to gender, "race" and social class although we need to avoid regarding these simply as "demographic variables." On the contrary, we need to consider the diverse ways in which young people themselves construct the meanings of those differences, and how they are defined and mobilised in different social contexts. (1993b)

The developing mind is a complex variable in this equation, casting further doubt on the simplistic notion that children are uniformly moldable. First there is the problem of accurately representing a child's perspective without tainting it (consciously or unconsciously) with adult viewpoints. I ran into this several times when I tried to interview young (K–5) children.

The following is a transcript of part of my interview with a second-grade girl, Meredith, about TV shows she liked to watch. Her mother, Cheri, facilitated.

ME:	Why do you like *Full House*?
MEREDITH:	Just cuz it's good.
CHERI:	What's good about it?
MEREDITH:	I don't know.
CHERI:	You like all the characters?
MEREDITH:	Yeah.

I was, in effect, asking Meredith to translate her internal viewing experiences as a child into an adult characterization. It obviously didn't work. I had a similar problem with Jimmy, a sixth-grader. I was looking for his thoughts about the second-grade movie his sister had made.

ME:	Do you remember the movie now?
JIMMY:	Yeah.
ME:	Any favorite parts, or did you think it was good or bad?
JIMMY:	Pretty weird.
ME:	What was weird about it?
JIMMY:	Funny weird.

The children I worked with weren't comfortable expressing ideas verbally. They would more often than not answer questions with "I don't know" and laugh, embarrassed about being put in the formal "spotlight" of an interview—a temporary position of authority.

Kids clearly perceived the act of interviewing as an adult activity. Over time, I learned much more about them when I stepped into their worlds and did things with them. They would casually reveal their feelings and thoughts either in their actions or in natural conversations in the context of some activity.

Another potential problem in trying to understand children from an adult perspective, added Buckingham, is the temptation to overromanticize them.

> In the case of research on youth culture, for example, the attempt to identify oneself with the "other" has occasionally led to a romanticisation of forms of "resistance." . . . In the case of younger children, it is often hard for researchers (and their readers) to avoid a Wordsworthian marveling at children's innate wisdom and sophistication, or a vicarious identification with their anarchic—but nevertheless terribly cute—rejection of adult norms. The difficulty many adults experience in listening to children without patronizing them is a direct consequence of their own power. (1993b)

The point of this is that the quest to find and display a child's perspective or voice is not a straightforward or simple one.

Schooling the Developing Mind

Children are engaged in an intense process of social preparation through their schooling. Their formal education in this sense might be understood as a set of stages on the way to a more mature state of sensibility. But they are not without such sensibility, as Olson related, when they arrive to school. They have had years to accumulate home-bred common sense from the cultural environments they are raised in.

> Common sense is a set of socially shared, basic beliefs that are presupposed by ordinary perception, action and interaction and which may be used to explain, legitimate, rationalize, and so render comprehensible those actions and interactions to members of that social group(ing). (1987)

"School," Olson went on to explain, "actually enhances and broadens the infant level of common sense children carry into schooling allowing them to extend their very basic senses to a larger world, and it is this very common sense that permits education to occur" (1987).

But within structured learning environments, children are moving targets with wide ranges of learning ability and cultural upbringing. Schooling in this sense could be said to dull a child's existing common sense by trying to force it to conform to an institutional code.

Olson elaborated on how such a code—a school system's "sense"—can transform a child's culturally innate "sense," not necessarily for the better.

> Schooling enforces the reception of a new bureaucratic, majority view sense which often works against the culturally based senses children arrive to school with, and this explains at least in part why some children (particularly those from minority, or disadvantaged sub cultures) generally have so much trouble with education. (1987)

My composition lesson with Mrs. Beechwood's class demonstrated this tension to a degree. The "adult-minded" lesson I had planned stood in clear contrast to the undercurrent of the children's interest in seeing themselves. Their involvement in my lesson plan required a degree of restraint of their "un-adult" senses.

Mrs. Beechwood noticed this in an even more general way during the screening of the movie we made with her first-graders. To her, it was a matter of personality:

> After a whole year of teaching the kids, some of them become so dear to you and there are others . . . [sweetly grimacing] . . . there's just that personality, that every day. . . . And I think that the video . . . it brought them back to six-year-olds for me. You can see their sweetness. And someone like Willy, who tried your patience, by 9:02 you're ready to kill. And you see that big smile and that grin. And he really is a lovable kid, but you kind of lose that sometimes. But they became even more lovable in the video . . . that brought them back to being six-year-olds.

Personality in this sense was often seen as a threat to the institutional sensibility of the classroom.

Developing Minds and Television

Television, not unlike childish behavior, has generally been placed outside the circle of appropriate educational activities. The parents and teachers I interacted with at Corrigan Elementary tended to shield children from television as if too much exposure would stunt their development.

The media literacy movement grew out of this protective stance: to inoculate young viewers from the potential dangers of television. Ironically, this shielding could be one of the reasons why children so easily identify with television. They do not tend to perceive television as formal, demanding, or educational. Instead, they see it as enjoyable, easy, and entertaining.

"The 'hidden' message contained implicitly in all that they watch," explained Cedric Cullingford in his book *Children and Television,* "is that they should be amused, and that television is dedicated to the art of entertainment. For this reason children find it difficult to take television as seriously as critics, and difficult to attend to in the same way as a lesson in school" (1984).

In other words, children tend to perceive television as a form of recreation rather than work. This may be a blessing more than a problem. TV's disarming quality to kids may open up parts of them otherwise closed. Many young learners are intimidated and frustrated by traditional learning procedures, and television could be considered a comparatively more "friendly" instructional method to reach them.

As far as parents' worries about the dangers of television, Cullingford suggests that part of the problem is in underestimating children's ability to handle it.

> While there are some clear findings about the ways in which children respond to television, these are always in contrast to children's capabilities.... Instead of trying to make sense of the complex nature of children's response most of the generalisations have assumed that response is simple and that television is complex. (1984)

He expanded on this by pointing out how children differentiate between the real and the fantastic: "They know that the advertiser's job is to sell; they know that the deaths in a Western are all simulated. But the significant fact is that they do not care. They become indifferent to the juxtaposition" (1984). When I first witnessed this "indifference" in the children I worked with, I was concerned. I worried that they didn't "get it." Over time, I discovered it was more a sign of "ho-hum," bordering on the obvious for them. Why should they get excited about the fact that the world is round, gravity makes you fall, and TV is a business?

The point in this is, we don't have to dwell on the obvious. Children are not defenseless when it comes to television. This does not mean that they have all they need to deal with television, only that their resident senses can be called upon and put to work in an educational setting.

The Culture Project

We did just that in Mrs. Spencer's third-grade class with a movie called *The Culture Project.* Like most teachers I worked with, Mrs. Spencer was skeptical of television's educational value given its propensity for escapist entertainment. Nevertheless she agreed, due in part to pressure from her own students, to give it a try. The moviemaking experience was becoming a tradition of sorts at Corrigan and Mrs. Spencer was genuinely curious about the prospect of connecting television to a lesson plan.

We sat down and discussed her curricular goals and the moviemaking process. She identified very closely with social studies and, of all the subjects in her class, wanted to involve this unit in a movie. The kids had been working in groups on reports about different cultures of the world. We decided to use the movie to get kids more involved in their topics and underline the overriding curricular theme that when we learn about other cultures, we learn more about ourselves. She was hopeful that we might be able to do some sort of "more serious" genre like a documentary.

Based on my experiences with the earlier grades, I suggested we put the challenge to the kids and she agreed. The kids could not have been more excited about the challenge. They immediately came up with ideas in all directions, which is always a good sign because they were embracing their lesson plan as much as the movie. The challenge would be in coming up with a coherent story that somehow involved as many students as possible.

The first topic to deal with was genre. What kind of a movie should it be? I immediately coaxed them in the direction of nonfiction, suggesting we could make "mini-documentaries" out of their group reports. They moaned loudly in defiance.

Personally, I sided with Mrs. Spencer, because it made sense to work their nonfiction reports into a nonfiction genre. I also worried about the children's possible aversion to more "serious" television forms since everything I had done with them in the earlier grades was fiction and fantasy based. I felt a certain responsibility in expanding their moviemaking horizons in new directions, and this project seemed to present the opportunity.

But the kids were simply not buying it. They did not see a documentary when it came to this project. They saw a story leaning in the direction of magic and special effects, and really wanted a stake in this lesson plan. These children were questioning "the sculptors," and clearly were not responding in a claylike manner. They were using their voice.

Voice

Voice is a term often used in media production to refer to the viewpoint behind media content. Media messages are expressions of different shapes, sizes, purposes, and genres evolving from the motives of media-makers. Children, Michael Emme, writes, are notoriously absent from such expressions as a rule:

> In a society where images serve as cultural currency having your image "out there" and having control over it makes you visible . . . children are invisible in our culture. If you eliminate all of the adult-made images of children from the mass-media you are left with blanks. If you assume that children perceive the world differently than adults (a basic tenant of art educational theory) then it is fair to say that we have no idea what children look like culturally. With all of that in mind, one of the mandates of media . . . is to facilitate children's self-representation in the mass-media.[1]

Media can be said to give amplification to individual voices because they project voices beyond the limits of a physical voice. The students working on *The Culture Project* were very aware of the opportunity their movie offered them to be "heard."

This is not to say that the need for children to be heard should outweigh an educational activity or objective. This is simply an example of the force of their expressive interests that can be called upon when appropriate for learning activities.

The more I worked with young children in their classes, the more conscious I became of the divide between adult and children perspectives and how easy it was to overlook and underestimate a child's expressive desire.

In *Common Culture: Symbolic Work at Play in Everyday Cultures of the Young,* Paul Willis saw this desire as an ever-running stream in the everyday life of a child.

> There is a vibrant symbolic life and symbolic creativity in everyday life, everyday activity and everyday expression—even if it is sometimes invisible, looked down on or spurned. . . . Young people are all the time expressing or attempting to express something about their potential or actual *cultural significance.* This is the realm of living common culture. (1990)

In her book *Literacy in a Digital World: Teaching and Learning in the Age of Information,* Kathleen Tyner saw media production as a means to tap into this stream of cultural significance.

> Voice is a concept that transcends the vagaries of the image or even the politics of identity. Specifically, media production gives voice to students who are otherwise silenced in their schools and communities. It allows students to represent their experiences and their communities as cultural insiders,

instead of the incessant representation and misrepresentation of them by media producers outside their communities. (1998)

David Gauntlett demonstrated this in his media production research with young school children. He observed children from several different schools as they made videos about their local environments and concluded,

> Young video-makers are undeniably reflexive and media-literate producers of original material, amply demonstrating the thesis (Buckingham, 1993; Gauntlett, 1995a) that children have an understanding of media which is much more sophisticated and creative than media researchers have traditionally assumed. (1996)

This was certainly an accurate description of my own realization as I worked with K–12 kids. First, I came to realize they knew much more about media than they let on, and second, they were more deeply involved in the content of it than I had imagined.

I realized this after I had tried for years to enlighten them about media structures they already knew and didn't appear to care about. It wasn't that they were indifferent to the structures, it was more that they were involved with media in their own way—a way that was not immediately clear to me. The more I interviewed kids and their families, the more I realized it had to do with their intellectual involvement and negotiation with content. They were feeding on content like bees on pollen—and in some complicated way, using it to play with their identities.

Kids grow physically every day. We don't see it happening day by day, but over time—like a time-elapsed movie—they mature. Their intellectual development could be thought of in the same way. Their interaction with media (both reading it and writing it) could be used as a means to capture key moments of their social, cultural, and psychological development.

Thinking back to my first three years of making movies with K–2 classes, I realized there were at least three ways children tended to express themselves when it came to our videomaking activities:

- Watching, interacting, and identifying with TV programming
- Being seen and heard by others
- Seeing and hearing themselves

I used these observations to begin formulating a sense of children's voice through media. In many ways, these expressive activities reminded me of a sociological school of thought developed by Herbert Blumer (1969) called symbolic interactionism. Blumer described human behavior from the standpoint of how individuals used their interactions with each other to formulate their identities. There were three basic premises to symbolic interactionism:

1. Humans act toward things on the basis of the meanings the things have for them.
2. The meaning of such things is derived from the social interaction one has with others.
3. These meanings are handled in and modified through an interpretive process used by the person in dealing with the things he or she encounters.

One of the most applicable aspects of this theory to my activities was in the way it accounted for the complicated dynamics of developing minds. This was one of the features of working with the kids that I needed help understanding. It was important to see children not as fixed and formed entities but rather as continually moving, changing, and negotiating social beings. It seemed to me that any framework to explain children had to be as dynamic as they were.

Applying symbolic interactionism to children and media meant seeing television as an observation post for the process of identity formation. Blumer's premises describe individually observable face-to-face social interactions within the contexts of everyday life. Interactions involving children and media are comparatively more transparent and observable. In this light, media can be seen as a means of glimpsing into the identity formation of children. In effect, observing their interactions with television, both watching and producing, allows us to capture snapshots of their identities in the making.

The classroom is an appropriate place for children to "play with" notions of their identity and the complicated ideas about life that go with it. The developing mind and voice of the child are, at best, works in progress. Children don't have it all figured out, so they need safe spaces to play, air out perspectives, and rehearse their identities. Teachers are perfectly situated to weigh in on this process, particularly when children need them, and televisions and cameras can be instruments of such experience.

Brainstorming *The Culture Project*

Despite our worries, Mrs. Spencer and I decided to go with the kids' fantasy approach to the project because they were so vehement about it. But we reminded them that the primary purpose behind it was not fantasy but rather sharing perspectives about other cultures. Hopefully, they would not lose sight of the lesson behind the movie exercise.

Many of the children in Mrs. Spencer's class had already made movies in their earlier classes, so they were confident and took charge.

The basic curricular objective was to get them as involved as possible with their study of world cultures, thus the name of the project: *The Culture Project.* They had already begun research in small groups on specific world cultures resulting in one- to two-page summaries of countries around the globe: Mexico, France, China, Japan, and Russia. Mrs. Spencer would have loved it if they could have read the reports out loud like reporters or documentary filmmakers. But this group had other ideas, *very different* ideas.

The first step was for them to answer the question: *What should we do a movie about?* They brainstormed:

- Different cultures from around the world
- Today Show
- Letterman
- Late Show
- Weird Science
- Weird Cultures
- Coyote Talk
- More Than One Host
- Funny (Joke Jox)
- Magic Tori Gate
- Time Machine
- Watches in Sync
- Car without Wheels in a Box (Time Machine)

The progression from reality to fantasy was quick and irreversible. The idea of a time machine was the turning point. Whatever we did had to have a time machine in it. Even though the genre was shifting out of the realm of "adult preference," the children kept the academic motive of the project in sight, as indicated by the incorporation of "science," "culture," and "Tori Gate," in their brainstorming activities.

But these young creators were taking authorship over the project and lending their voice to—or perhaps, more accurately, "forcing" it upon—the class project. This was, at least in part, an example of how children brought their TV viewing experiences into their own worlds.

Voice: Through Watching, Interacting, and Identifying with TV Programming

In the case of brainstorming for their *Culture Project* movie, Mrs. Spencer's students were indicating their identification with programming in the ideas they were putting forth: everything from adult shows (*Today Show, Letterman*) to children's TV shows and movies (*Weird Science, Coyote Talk*).

This was very common in my experiences with them, and it tended to worry adults I worked with, at least at first. It could be considered a form of mimicking—mimicking programs and ideas they liked in the programming they watched, and projecting them in their own creations. Some might argue their imitation of existing media could be considered a re-presentation of adult portrayals, therefore circumventing their own "pure" voices. The same could be said of words in the formation of written language—some level of imitation, at least to start, is inevitable and part of the process of getting airborne with their own voice.

Children's use of existing programming structures can not only provide a template for eventually original expressions, but it can also serve as an opportunity for articulation of cultural issues. Teachers can take advantage of lessons in the making that come with answering questions like "Why is there so much violence in stories?" or "Does violence have to be part of every story?"

The point here is that if we consider their interaction and identification with existing programming, we have an opportunity to converse with them as they consider ideas that come to them from media sources. We can question them, accept them, provide alternatives, and exercise their critical thinking about issues that arise as they contemplate media messages.

The kids creating *The Culture Project* were drawing from existing structures of stories they had seen in various media, and blending them with their social studies project, despite the gentle reluctance they were getting from the teacher about straying from the nonfiction quality of most traditional social studies activities.

Scripting *The Culture Project*

The children wanted to make a movie that would take them to the countries they were writing stories about. The setting would be a third-grade class where a special club of students who called themselves "The Culture Project" got together to do culture projects, like students in a science club would do science projects. They worked their time machine idea into the club and began to deal with some story issues such as characters and plot.

SHOW STRUCTURE OUTLINE
WORKING TITLE: "CULTURE PROJECT"

I. OPEN / MUSIC / TITLES
II. CULTURE PROJECT DOES THEIR TIME MACHINE THING
III. PRESS THEIR BUTTONS AND WE TRAVEL
IV. PROBLEMS (MECHANICAL, AND BAD GUY)

SOLUTION/CONCLUSION/CREDITS

This was enough to get a basic start. After the class left, Mrs. Spencer and I conferred on the results of the story meeting with the kids. Although she was impressed with the energy and creativity of the exercise, she still wondered a little about the academic merits of the project and whether it would teach them anything. Could something so fun for the kids teach them something or would it have to be "written off" as a recreational experience?

Voice: Being Seen and Heard by Others

The next big challenge for the young filmmakers was to come up with characters and a story: Who would play which roles? How many characters should we have? How would we equitably divide roles, both performing and crew?

This is a question that always seemed to strike the heart of the notion of a child's voice. This is because it was all about them considering how they would be seen and/or heard, by others. In a symbolic interactionist sense, considering how others perceive you is not all that different from considering how you perceive yourself. But for the sake of simplicity, we will deal with "others" and "self" separately.

Since they were putting together something specially designed for public presentation, they were keenly aware of the opportunity they had as creators to control how they were perceived. Naturally, some wanted to be seen in the limelight and vied for leading roles. Others preferred background roles because they were uncomfortable being seen "up close." They wanted their voice to be "heard" through their work behind the scenes.

The moment of truth in being seen and heard by others was the public screening of the finished movie. Second-grade teacher Diana Green shared her memory of her first screening, how the power of their recognition seemed to last well beyond screening:

> You could just see their chests puffing up, and they were so proud of themselves, and proud of each other—which is another major step. They can be proud of themselves, but when they're proud of each other, that's really the kind of thing we're looking for, the social awareness . . . and it lasted all year. . . . People would stop in the hallway and say, "We loved your movie!" You could just see them feeling so good about it and feeling so good about themselves, and about their class.

Having others see and hear their "voices" confirmed and strengthened the children's sense of personal significance and self-esteem.

Voice: As Seeing and Hearing One's Self

The children were also watching themselves on the screen and feeding their individual perceptions of themselves as if they were looking in a mirror. This was one of the first across-the-board realizations I had when I did movies with children. There was always a noticeable spike in participants' self-esteem levels when they watched themselves on the screen.

Diana Green certainly noticed this with her students:

> The overall effect that I had never anticipated . . . the raising of the self-esteem . . . on the class as a whole and on individual children.

Just like the time I first turned the video camera on Mrs. Beechwood's first-grade class, their experience in seeing themselves and their accomplishments on the screen seemed to tap into a reservoir of incredible energy. In the end, there was no more clear indicator of how important it was for children to "see themselves" than "the credits" of their movies. Seeing their name was as important to the children, sometimes even more, as seeing their face on camera. Diana Green called it one of the most important parts of the movie for her students:

> When you get done watching a movie and the credits come on, everyone gets up and walks out. . . . But when their credits came on, the kids were almost ready to stop paying attention to the movie, then they started looking and going [mimics their gesture of awe].
>
> And the fact that this was in print, I think, seemed to have some kind of effect on the kids. They were very impressed with that, and that sort of seemed to equalize everybody, that they were all important, that they made that movie.

Credits are normally not child's fare, but they enjoyed seeing their own names and this may have solidified their understanding of authorship—in other words, that media messages are made or authored by particular people. But most importantly, it meant that they were being officially noticed and certified as creators. Their "voice" was publicly legitimized with the credits.

There was also a particular power in the knowledge that they had as creators. They held all the secrets—the camera tricks, the real stories behind the action, the special effects and editing techniques. They knew something others—in most cases, their parents, loved ones, and friends—didn't know.

Division of Labor

Mrs. Spencer's students decided that everyone in the class should be involved in some type of performance, even if they were just seen on the screen as part of the class. That included Michael, a physically handicapped child who could not speak. Two of the students would have primary roles as the protagonists of the story, but all of them would have the chance to be seen and heard as performers, including Mrs. Spencer. There would also be an opportunity for children to direct production activities, including lighting, microphone holding, and camera work.

They quickly hammered out a story outline. It would serve as our shooting plan for the production phase.

Movie Title: The Culture Project

Concept and Theme: We understand ourselves and our own cultures better when we learn to know and appreciate other cultures.

Setting: A third-grade class attending a yet-to-be-named elementary school in a yet-to-be-named city. As we get into the movie, we visit settings of other cultures by inserting pictures of those cultures (using the electronic effect of chroma key) behind our characters who are "trapped" in these cultures.

Characters: *Class members*—Everyone in the class is a character, and each performs a different role in the story. Most people in the class including the teacher (or teachers!) are mysteriously and quite mistakenly transported to other cultures around the world when a science project goes wild. Those who are transported represent the cultures that they have landed in while they are in the process of being rescued and brought back to the class.

"Brains" or "Science Kids"——Two girls who are good friends. One of them has invented a time machine of sorts that allows a whole class to be transported through a time and space wave to distant lands and cultures in only one millisecond. The other, not as "brainy," is a loyal assistant.

Bullies or "Bad Kids"—Two boys that deliberately thwart all decent class activities by causing problems in the class.

Problem: Something goes wrong with the time and space transportation device and it ends up sending all the kids in the class (including the teacher!) to different parts of the globe. The brain believes her assistant has caused the problem and this creates tension. The brainy duo has to work together and find a way to fix the problem before the principal discovers their teacher and classmates are missing.

Solution: One by one they make contact with each of the missing transports in their cultures and we get to know the cultures that they are trapped in. In the end, they all realize that they have not only learned something about other cultures, but also in getting to know those cultures, they have learned more about themselves and their own cultures.

Final Credits: Highlights of the behind-the-scenes shooting of the movie.

Producing *The Culture Project*

The movie was shot in one-hour sessions over a two-week period and it culminated with a half-day visit to the television studio at the university where I work. This is where we photographed the "lost students" in front of backgrounds that matched the countries and cultures they had been accidentally displaced to. The last scene to shoot was the Pacific Ocean, the place where the Mrs. Spencer had landed. We used an electronic special effect (sometimes called the weather map effect) where the subject stands in front of a green screen and we substitute an appropriate background. In Mrs. Spencer's case, we used video of an oceanlike body of water—to go over the green. The kids had a lots of fun with that.

Once all the video was shot, I edited it together for them. At the time, we did not have accessible editing technology like we do now, so I had to use university equipment. All of the editing that I did (and much more) can now be accomplished in just about any school using standard computer equipment.

We screened the movie and everyone loved it. It was a resounding success for the students in the class. After the excitement died down a bit, I sat down to talk with Mrs. Spencer about the experience. She was impressed with what the kids were able to accomplish, despite her original concerns:

> I really developed an appreciation for people who make movies. . . . It just was incredible to me how much you have to think ahead and plan. . . . You can't just fall . . . you know, the time machine and the magically appearing . . . I mean, are they going to get too far from learning about other cultures? But when the movie was done and when I saw it . . . it worked! It all came together nicely, everybody learned something from it, and they learned about social studies, about other cultures, how you can understand yourself better when you learn about other people—even with the fantasy aspect of it worked in!

When a Child's Voice Is Heard

The children had surprised Mrs. Spencer and surprised me too. Not only did they work harder than they would have had there *not* been a movie, but they also demonstrated uncanny visual sophistication in creating the project and solving problems, both creative and technical. I was slowly realizing that this rising generation was more visually acute than the one I was working with at the college level.

I was beginning to think that maybe I should be teaching my filmmaking students at an earlier age! Mrs. Spencer had learned a lot about her students too and couldn't help but walk away from this experience with a new appreciation for the capabilities of children. In this case, our experience of moviemaking was connecting teacher and students in fresh and exciting ways, worthy of exploring further.

Mrs. Beechwood perhaps explained this connection best when she talked about what she took from her first-grade moviemaking experience:

> Some of the things, when they weren't doing what they were supposed to, they looked sweet and cute. Harry and Willy sticking their face in the camera and doing the bunny ears . . . Jamal laying on the floor—and he looks charming. Those things look cute, but when you're in the classroom and trying to get them to settle down, those are the frustration. . . . And . . . you don't hear the frustration in my voice, "Will you please get in line!? I told you . . . get in line. Get in line! Get in line! Get over here!" You don't hear that . . . you just hear that sweet music. They're just kids, but I lose sight of that sometimes, because I'm worried for their safety. I'm worried that I have to teach them something. So you've got to pay attention. You've got to get in line.

Yes, they are just kids and they are desperate to be seen and heard as they try to make sense of their lives and the world around them. But hopefully everyone will agree, after all this, that there is one thing they clearly are not: They are *not* clay.

Note

1. Summarizing an article on photography and cultural invisibility, he (Michael Emme) had published in 1995, in the *NASCAD Papers*, a biannual publication of the art ed program at the Nova Scotia College of Art and Design.

CHAPTER 4

THE MESSAGE IS THE MESSAGE
Media Literacy in a Visual Age

> The effects of literacy on intellectual and social change are not straightforward.... It is misleading to think of literacy in terms of consequences. What matters is what people do with literacy, not what literacy does to people.
>
> —David Olson and Nancy Torrance, *Literacy and Orality* (1991)

IT WAS PROVING to be a particularly trying stretch as Elizabeth Redfield's third-grade class entered the final quarter of the school year. There had been a wave of school shootings across the nation over the past year and if that wasn't enough, "local terrorists" were doing their best to exploit the fear in the air by phoning in daily bomb threats to the school district.

Ms. Redfield was not feeling any particular joy—let alone creativity—when I asked her if she'd like to do a movie in her class. She was running out of explanations for her students. The bomb threats were becoming so routine that teachers and students were scheduling evacuations as regularly as math lessons.

She felt she owed them something, though—something more than a diversion from the confusion and extraordinary insecurity of the past two months. The bomb threats and school shootings were testing her own strength to come to school every day. It was a very scary time to be a teacher.

Between bomb threats and lesson plans, we talked about a movie experience that could address the very subject that was making the school year so difficult: fear. If someone had told me six years earlier that we'd be doing a third-grade movie about "fear," I likely would have been perplexed. What did fear have to do with media literacy?

Media literacy had once been the primary motive behind my bringing cameras into K–12 classes, but every year I worked with the students, it seemed I was straying further and further away from media literacy.

Media Literacy

My aim, as I saw it, was to arm them with critical viewing skills so that they could control the media images rather than the other way around.

Media literacy was a very sensible research agenda for me, I thought, having been a professional in the TV business and then an educator of media professionals. I shared a view of media literacy—with a focus on educating critical viewers—very similar to Ibrahim Hefzallah, who wrote,

> A critical viewer is a person who knows how to observe, and how to evaluate what is observed—skills needed to master the medium of television. Those skills should be practiced by parents and adults, and taught to children. And the best way of teaching those skills is by setting up models of critical viewers children can imitate. (1987)

I was finding rather quickly and consistently that this was a *painfully* easy and unimaginative objective. This is certainly not to dismiss the value of media literacy, especially in a world where media play such a vital role in the everyday society. But the more I worked with K–12 kids, the more narrow and patronizing I realized such a view of media literacy was. It didn't let the magic or wonder of moving-image media take root, rather advocating a notion of "proper" and "healthy" media consumption.

David Gauntlett found something of the same in his videomaking experiences with school kids:

> Children demonstrated a high level of media literacy in all age groups.... The study also shows powerfully that a methodology which avoids the patronising, positivistic stance of the psychology-based effects tradition and allows children to show their intelligence and discretion in relation to the media, can transform the kind of conclusions which must be drawn. (1996)

I was certainly finding this true in my own work with the children. The more I connected our videomaking experiences to the interests and needs of children and teachers—above and beyond the objective of making them media literate—the more successful our media-making activities were.

More and more, I learned that children were quite adept at videomaking. They picked it up rather intuitively and effortlessly connected it to learning activities they were involved in. It seemed to make what they were learning more relevant. Thus, the media literacy I was doing was expanding beyond the specific subject of media into the general subjects of traditional learning.

David Olson conjured up a larger picture around the aims of traditional print literacy that offers useful parallels for media literacy:

> The common goal is to determine all that is involved in our being and becoming literate. . . . We must go, it seems, back to the beginning. . . . Writing not only helps us remember what was thought and said but also invites us to see what was thought and said in a new way. It is a cliché to say that there is more to writing than the abc's and more to literacy than the ability to decode words and sentences. Capturing that "more" is the problem. (1994)

In the very same way, there is without a doubt "more" to media literacy than inoculating helpless viewers.

Monsters

Amidst the troubling events of the school year, Mrs. Redfield and I sat down to talk about the motives behind the third-grade movie project—in particular, the "fear" issue.

She was truly perplexed on how to address the bomb threats, the reasons why they happened, and also why there were such things as school shootings. These were all questions her students were asking and though she felt she owed them a straight answer, she really didn't have a good answer. She simply did not know how people (even children) could do such things. She needed an alternative means of talking to the children about the problems and reassuring them that they would be all right. She also felt it her duty to provide an explanation as to why these events were happening. Maybe a movie story could serve as a figurative exchange of sorts, a way of talking to her students on a gentler, more symbolic level.

With this in mind, we decided to create a story about the bomb threats but to use the metaphor of a monster to represent it. This would allow both students and teacher to step back from the "real" issue of bombs and project their feelings and creative energy into a story that could absorb their anxieties and provide a place for them to vent their fears and confusion. In essence, we were using the medium of television to help solve a problem that traditional means could not.

After we came up with this plan, I wasn't sure we'd even be able to explain it to the kids, let alone have them participate in it and create a movie story. But surprisingly they did, and reasonably effortlessly. They appeared relieved by the buffer the movie project was giving them. They were keenly aware that we were distancing them from actual, current events and seemed to appreciate it. They were also keenly aware of the connection between the story and the bomb threats.

They came up with the concept of a "monster threat" to parallel a bomb threat. A monster threat was basically the same as a bomb threat, except that it was pure fantasy to the children, whereas bombs were real.

After presenting the challenge to the third-graders, I stood at the blackboard and asked them to work out the plot of our movie so we could get started shooting as soon as possible. They responded with great ease and generated a story outline.

Monster Threat Story Outline

Erin's voice talks to us while she is cautiously walking down a dark school hallway facing a fear. She tells us the story of what events led up to this moment. Flashback to:

1. Erin and Michelle going to school to find out from the teaching assistant, Mrs. Goodfriend, that there has been a monster threat called into the school. Mrs. Goodfriend explains what a monster threat is.
2. The girls joyfully return home and end up at their neighborhood playground or "secret fort" in the woods, where they discuss their monster fears.
3. In school the next day, Evan and his bully bunch tease and intimidate the girls at their lockers.
4. Later in their classroom, Mrs. Merryweather helps fearful children confront their monster fears.
5. Children are sent home by the voice-over loudspeaker—another monster threat.
6. While at home watching television, Erin and Michelle begin to get bored of monster threats.
7. The next day at school the girls try to avoid Evan and the gang at their lockers but are thwarted by the bullies outsmarting them. They are becoming even more scared and frustrated.
8. Later in their class, Mrs. Merryweather tells them that "the monster"—although something we should take seriously (not ignore)—is not something that should control us. We must eventually confront "monsters" by being courageous and facing them, perhaps even

outsmarting them. This particular monster wants us to stop coming to school. This monster wants to defeat our will to learn and grow educationally. She suggests this might be because the monster is actually afraid of people who want to learn. She suggests that you can often discover what monsters fear most by observing the means a particular monster uses to frighten you.
9. Next day at the door of the school, Mrs. Goodfriend confirms that there has been another monster threat, and everyone has to wait for the monster search to be completed. The girls are so tired of this by now that they are mad. Michelle says she is going to go to her locker another way to avoid Evan and the bullies. Erin says she is not going to be intimidated anymore. Mrs. Goodfriend says students are cleared to go in the building now.

We are back to the present now where we began the movie.

Erin turns the corner to face her monster—Evan. She confronts him but he does not back down. All of a sudden, Erin looks behind Evan in fear. It's a real monster! Evan runs away in total panic. The "monster" removes its mask and we see it is actually Mrs. Merryweather. She tells Erin that sometimes you have to think like a monster to beat that monster. She and Erin laugh and go to class.

Erin's voice reminds us that even though Evan was not a problem anymore and the monster threats eventually went away, there would be more fears and monsters to face, but she would continue to resist them wherever they might pop up.

Mediums and Messages

"The medium is the message" because it is the medium that shapes and controls the scale and form of human association and action. The content or uses of such media are as diverse as they are ineffectual in shaping the form of human association. Indeed, it is only too typical that the "content" of any medium blinds us to the character of the medium.

—Marshall McLuhan, 1964

When I teach critical perspectives of television to college students, I warn them that they will never watch television the same way. This is because they become aware of illusions they once thought were real. Becoming aware of the formal properties and procedures of television aggravates carefree viewing because it brings what were once invisible structures "into the light."

The "deconstruction movement" in traditional print literature posed a similar challenge to book readers. Critically deconstructing literature involves seeing a literary work as a construction of an inherently biased author involved in selling a message to a publisher. There is no way to strip or disassociate the author's history and biases from the "truths" presented in his or her words. Teaching readers how to deconstruct literature *before* consuming it is not meant to destroy the literature or to take the fun, mystique, and value out of it. Instead, it is meant as a platform for a wider frame from which the reader can interpret the work.

In motion picture circles, viewers who are taught to deconstruct television literature become keenly aware of structures once invisible to them, and usually can't help but share the excitement of their newfound vision with whomever will listen. This newfound critical awareness has been known to promote high levels of chatter during films and television shows, ultimately putting stress on friendships. Of course, McLuhan never told us this. And he also didn't tell us what to do once we were critically enlightened about the true character of media.

Did it mean we would become blind to content in the way we were once blind to a medium's character? Could we ever enjoy content again, once we knew its secrets? I assure my students that it doesn't take long to grow out of their content blindness and redevelop the ability to slip into a message without deconstructing it first. It would just take a little time and a little concentration.

I realized in my literacy work with K–12 students that I was not contemplating a similar "exit strategy" in their media enlightenment. Once I had sufficiently demonstrated to them the ways in which "the medium was the message," they wanted to return, like the college students, to their earlier state of pure enjoyment where *the message was the message.* They just wanted to enjoy the stories again.

In the end, McLuhan was right about mediums and messages, but like the media literacy movement, his indelible adage "the medium is the message" didn't go beyond the enlightenment factor. What were we to do with the enlightened once they were literate? Simply being literate is, I've found, just the beginning. The end appears to have more to do with getting back to the content—in McLuhan's terms, *the message.*

In the monster threat project, we were not dealing with the media component as our "subject"; we were dealing with it more as an educational tool driven by a purposeful content objective—handling "fear." We were moving past the medium to the message.

What the Camera Sees and Doesn't See

Though it looked like an elementary school to the eye, Mrs. Redfield's third-graders had discovered it looked a little different through the camera lens. Their script called for a shot of a fictional elementary school to open the movie, so they figured they could easily acquire a picture of their own school to use for the story.

From the outside, it looked pretty much like what it was—a typical American elementary school, not all that different than the one I went to years ago. Over the past century, its eclectically arranged red-brick components had acquired the capacity to hold seven solid grades (K–6), up to eight hundred children. School colors, royal blue on gold, adorned a rather unpretentious sign at the front corner of the building: "Corrigan Elementary School: Kids Are Great, Aren't They?"

We weren't sure of the best angle to videotape the sign, so we crossed the street to allow for a longer shot perspective of the school. The student assigned to the video camera seemed confused about what he saw in the viewfinder.

"It's OK," he reported to his twenty or so crewmates crowded around him. "But it looks a little small."

When I looked through the viewfinder, I saw not only a building that appeared far smaller than it was in real life, but what little could be seen looked more like a warehouse than a school. It certainly didn't measure up to the expectations the class had for a shot of a typical elementary school, despite the fact that it actually was.

It may have had something to do with the fact that there was not enough brick surface in view. The angle of the sign was certainly not thought out for a camera view. Its background was a yellow cement section of building, perhaps built in a 1970s or 1980s cost-effective add-on. The sign was positioned to be seen by cars passing either of two streets that intersected in front of the sign. Passing motorists would already know what the school looked like and wouldn't require it to be perfectly positioned in the visible background behind the sign. Their three-dimensional perspective would place the sign in context with the school.

Our movie audience would have no such context. We would need to construct a sense of the sign and the school for them with our two-dimensional representation. We searched for a better vantage point, arriving at a far corner of the building, brick-side. The challenge then became to somehow find a view through a bountiful row of maples that obstructed the school's "best side," not to mention the sign would be unreadable at this distance. The supposedly simple task of gathering an establishing shot, a shot we all clearly saw "in our heads," was turning out to be a time-consuming and near-impossible task.

Through it all, there was an important realization. Students were immersed in the process of seeing—capturing an image of what they saw in their minds and using a recording device designed to help others share in the same perception. What became clear was that the image in their heads was a collection of visual impressions gathered over time from different perspectives, but no single one of them wholly revealing. They were trying to fit this complicated impression of their school into a single, simple shot and realizing that it wasn't possible. It wasn't visible in a two-dimensional way.

After several more attempted setups, we recorded what little the camera could see and returned to the classroom. Total time for one simple shot of a school building: thirty-five minutes. Such is life in the business of photography. Nothing ever looks quite the same through a camera as it does outside of it. But the camera, as well as the process of using it, reveals much that is invisible, or taken for granted, in everyday three-dimensional life.

The experience of framing the shot of the school sign revealed how the world from the camera is different than the world outside of it. The camera changes things—it can limit or hide what we see in real life *or* reveal things that would not otherwise be seen. In the case of this video adventure, it revealed the difference between the two-dimensional—video images—and the three-dimensional—what we perceive with our eyes and brain.

At the simplest of levels, the students were aware of the manipulation they were engaged in as well as the wider implications of manipulation at play in the media they consumed. This critical level of understanding media is something that media literacy has traditionally focused on, almost as an ultimate objective: to make viewers, especially children, critically aware of the manipulation behind seemingly "real" media messages.

But as Kathleen Tyner suggests, engagement in the conspiracy of media manipulation, from a media production standpoint, provides more opportunity for children than simply making them critically aware consumers. Being aware is but a catalyst to a larger set of possibilities.

> Students already know the codes and conventions of media, especially of television, but they may not have the vocabulary to articulate that knowledge. Together, teachers and students can learn vocabularies: key concepts; and the economic, social, cultural and historical contexts of media production by exploring a formal and structured course of media studies that marries hands-on video production with building skill in media analysis.... Since the process of hands-on student productions lends itself to work that is interdisciplinary, group-oriented, and inquiry-based, it can support the goals of constructivist classrooms to lead students into areas of sophisticated, critical thought. (1994)

The difficulty of finding a shot to match their three-dimensional mind's-eye image of the school illuminated certain things to the children. First, their image of the school was a complex composite built from many different viewpoints and impressions, from experiences and memories over the years. When they saw the simple view of the sign through the viewfinder with no evidence of the qualities they associated with the school, they realized that their impression was based on experience more than visual reality. This was not just an exercise in framing a shot. It was a lesson in culture—children's taken-for-granted notions of the world around them—in this case, their personal impressions and definitions of school.

Duplicating such an impression would require some work on their part in enhancing the incomplete visual reality, perhaps even relocating the sign to an area that provided a more evocative background. In short, their impression of the school would have to be discussed, agreed upon, and visually translated (rather than visually transcribed). They would need to use the camera as an instrument of their artistic and cultural expression, because it was incapable of such expression on its own.

This is the point at which media making can become an instrument of more complex learning, when it is harnessed to educational causes. Media, as David Buckingham relates, can be a tool of cultural participation:

> Media Education is not confined to analyzing the media—much less to some rationalistic notion of "critical viewing skills." On the contrary, it seeks to encourage young people's critical participation as cultural producers in their own right.... It is essential that the curriculum should equip young people to become actively involved in the media culture that surrounds them.... Such a curriculum would encourage children to have high expectations of the media themselves.... The age in which we could hope to protect children from [the wider adult] world is passing. We must have courage to prepare them to deal with it, to understand it, and to become active participants in their own right. (2000)

The "medium" has long been the focus of media literacy's message. Buckingham and Tyner advocate instead for a more content-creation, production-centered media literacy. The outcome of such an active approach to media study is far-reaching because it transcends media and allows students to connect media processes to broader literacy objectives of traditional curricula. In a McLuhan sense, instead of "the medium being the focal point of the media literacy message," media literacy educators might consider the notion that "the message is the message." But this would not be just any message. It would be the message produced by students and how such involvement in media not only illuminates media processes and consumption but also stimulates thinking beyond media to issues and ideas in students' class work.

Video Prism

In the case of the monster threat, students were becoming media literate by participating in media making. But they were also engaged in a learning objective that transcended media literacy. They were using video to solve a problem in their classroom.

In this sense, the camera they were using functioned like a prism. It "illuminated" its subjects in alternative lights. It allowed users to see and portray aspects of life in ways that might not be seen otherwise. But there was also utility in the converse notion of videomaking: It could also make the normally visible features of its subjects *invisible*.

In the same way that a glass prism can illuminate the colors within everyday "white" light, splitting it into a rainbow of violet, blue, green, yellow, orange, and red, videomaking can selectively illuminate an array of aspects within everyday ideas and imagery. From this standpoint, video can

- distort subjects
- enhance subjects
- capture truths
- alter or question truths
- reveal what we might otherwise miss
- conceal what we might otherwise see

Video Can Distort

Using the genres associated with moving-image media (for instance, film, documentary, TV show, news, music video), K–12 teachers and I demonstrated that we could enliven subjects by essentially putting them through different "filters" of presentation.

For instance, when we used a news show format to showcase students' research skills, the "distortion" of the news genre made the research process more relevant to the students. It connected the lesson plan to something they understood from their television-watching experiences.

The same thing happened when we "distorted" a poetry lesson by filtering it through the music video genre. The action-oriented experience of making a music video distorted the poetry lesson plan in such a way that it suddenly seemed more relevant to the kids: not a lesson, but rather part of an activity they were familiar with.

Moving-image genres are tools of discourse that add strength and versatility to a student's means of expressing his or her ideas and perspectives. They offer alternative presentations of ideas that students might more effectively identify with.

Video Can Enhance

In scripting a story, videomakers choose what will and will not be in the story. This grants power to the storyteller to enhance what he or she wants to call attention to, perhaps even embellish, in a story to make a point.

When a fourth-grade teacher decided to hold a camping trip that her late colleague had always celebrated, she wanted to use video to capture the pure and delightful playfulness of children in nature. The video enhanced this message by focusing on such playfulness, as well as a playful song, called "Smile!" to back it up.

Such an enhancement factor can be seen when documenting any shared experience and watching it (reliving the experience) with the children some time after the experience. The experience and its significance to the kids have always been enhanced by video documentation in my experiences.

Video Can Capture Truth

Just like the experience of photographing the school sign captured a certain truth of its obscure surroundings, videomaking can capture certain truths that might not otherwise be seen.

First-grade teacher Mrs. Beechwood always talks about how video helped her to see first-graders for who they really were—little children trying to figure out the world around them. It helped her appreciate this truth when many of her daily routines designed to keep order tended to stifle such things. The video she made with her students helped her to appreciate acts that would normally be construed as mildly disruptive as innocent and unique. The distance that video afforded her view of the children revealed this truth.

When sixth-grade teacher Danielle Ippolito unleashed her video autobiography project called *All about Me,* she had to deal with certain aspects of truth that did not always fit into class protocol: language, humor, immature expressions. Even though she had given them the license to express themselves—certainly with some level of negotiation (not a lot)—she was surprised at the respect the students had for appropriately restraining their "free expressions," especially when it came to respecting each other and connecting to the collective spirit of the project. The "truths" of the autobiographies, however borderline, were constructive and revealing as a class experience. The video lens offered an alternative stage for students to present their "true" identities—identities that normal class activities tended to suppress.

Video Can Alter or Question Truth

Many times teachers used the prismatic quality of video to question or alter what normally might be perceived as "truth" and put that truth in a different light. This could encourage students to question or reconsider the truth, or simply to contemplate the critical implications of the truth.

Diana Green did this when she cast roles for her second-grade movies. She preferred to make a young woman "the hero" of her movies when she could, because she wanted the girls in her class to expect the very best of themselves. Other teachers also used casting as a means to alter the truth for a positive cause.

Sometimes they would portray strong students as weak—mostly because they could handle the irony of such a portrayal—and vice versa. Others placed shy people in more vocal roles.

In a middle school project, students "played" with the idea of gender expectations by staging a basketball game between a group of boys and girls. The girls bet the boys their shoes—thus, the name of the film: *Shoe Jam*—and they won handily 21–0. The interesting thing was that the girls in the class had no idea how such a story could be told—boys losing to girls in basketball. But the power of video storytelling to alter the truth not only shocked and inspired them, but it also got them thinking about taken-for-granted notions in their worlds that might not be best taken for granted.

Video Can Help Us See Things We Might Otherwise Miss

One of Mrs. Beechwood's greatest motivations for doing a video project was to capture the growth (both physical and intellectual) of her first-graders throughout the year. She loved to share the changes with the children and call attention to their growth and accomplishments. When her students and their parents watched the video at the end of the year, they were always impressed with how much they had grown.

The "repeat factor" of video was another feature of the videomaking experience that created extra visibility. Once something was caught on tape, it became timeless and magic, capable of being repeated over and over, each time with different realizations. Teachers could show their videos to future classes and reap the rewards from them even though past students had created and starred in them.

Videomaking also archived the past—old playgrounds, old classrooms, old teachers—and preserved them for future viewing and appreciation.

Video Can Conceal What We Might Otherwise See

We were using video to "conceal" reality, or at least soften its hard edges, in Mrs. Redfield's class. This was to refract the issue of fear and terrorism to a more manageable form that third-graders and their teacher could handle. In essence, we were concealing the truth in order to better handle the subject of fear with the third-graders.

We could also conceal our environs with a little ingenuity. Many times we would shoot a scene in the corner of a classroom that was supposed to look like a bedroom (in the script) rather than the classroom it actually was. We could pull this off if we skillfully concealed what was visible to viewers. This was accomplished using framing, focus, scenery, and props to create an illusion of a bedroom environment.

Using the Video Prism

What can video reveal in the "white light" of education that cannot be otherwise seen? Given the imagination of children, the sky is the limit. The prism is really where the medium gives way to the message. The message comes from the kids' spirits, the curriculum, the curiosity, challenges, and fellowship of education.

How do underfunded, and often unimaginative, school systems deal with such benefits of videomaking in curricula?

First they might try to see video as a tool of learning, rather than just a subject of learning. Children are already visually literate and bring much to the table as far as innate visual skills and certainly comfort in dealing with visual ideas and technology because they have grown up saturated in it.

Second, we have to look to the children, let them lead a little. The irony of media literacy in this day and age is that the teachers tend to be *visually illiterate* in comparison to the students. Teachers must therefore trust that their students can pick up on visual media-making experiences, even though they may be uncomfortable with the process themselves.

Third, educators need to exploit the novelty of television and videomaking and capitalize on this novelty by igniting kids' spirits through stimulating visual activities. In essence, video is for teachers bold of heart when it comes to learning. They do, in fact, need extra patience and determination (the kind that comes with good teachers) to pull video off. But even though there is novelty in video, it is important to realize it is no magic bullet. Clark and Salomon elaborate on this:

> Past research on media has shown quite clearly that no medium enhances learning more than any other medium regardless of learning task, learner traits, symbolic elements, curriculum content, or setting.... Any new technology is likely to teach better than its predecessors because it generally provides better prepared instructional materials and its novelty engages learners.
>
> While media are not causal factors in learning, they often provide the focus for curricular reform.... We should notice that new media such as computers *allow* for flexible and local construction of the conditions that facilitate skill cultivation, even though these materials might also be constructed in other ways. In this fashion, newer media serve as a *proxy* for the causal variables that influence learning and performance. (1986)

Conclusion

In considering a wider playing field for media literacy—at the very least, extending it into the production arena—it seems that Olson and Torrance's suggestion that "what matters is what people do with literacy, not what literacy does to people" is particularly appropriate to build upon.

A distant and passive exposure to media literacy will likely prove incomplete without an experiential reference point, for with such experience come meaningful goals. Olson and Torrance elaborate on such goals as they related to changes that evolved in response to reading and writing literacy:

> Literacy does not cause a new mode of thought, but having a written record may permit people to do something they could not do before—such as look back, study, re-interpret, and so on. Similarly, literacy does not cause social change, modernization, or industrialization. But being able to read and write may be vital to playing certain roles in an industrial society and completely irrelevant to other roles in a traditional society. Literacy is important for what it permits people to do—to achieve their goals or to bring new goals to view. (1991)

On top of the notion of a more active and participatory media literacy, Olson suggested long ago that there may be issues of educational form that developments in communications can pave the way for:

> The communications revolution necessitates a new, more broadly construed conception of the educational process. Perhaps the function of the new media is not primarily that of providing more effective means for conveying the kinds of information evolved in the last five hundred years of a book or literate culture, but rather that of using the new media as a means of exploring and representing our experience in ways that parallel those involved in that literate culture. In this sense, media are not to be considered exclusively as means to preset ends but rather as means for reconstructing those ends in the light of the media of expression and communications. (1974)

Mrs. Redfield was at a loss for how to deal with the issue of terrorism and the fear it provoked in her students and in herself. She used video to explore an alternative approach to this difficult subject and, in the end, it worked beautifully. The experience of writing the movie and of confronting fears through open discussion and abstract representation made her feel that she was doing something to strengthen her students and deal with the problem.

In this case, the headlines crept into the classroom, and video provided her the means to constructively engage in them and the fear they induced. In the end, it wasn't about what the class's newfound media literacy did to them. It truly was about what they did with their literacy. They battled and beat a monster.

CHAPTER 5

WRITING WITH CAMERAS

<u>SECOND-GRADE BOY IN WHITE HOODED SWEATSHIRT STRIDES ACROSS SNOWY FIELD AND STOPS TO PICK UP A PIECE OF TRASH, AND THEN LOOKS UP AT CAMERA.</u>

                      BOY
You know, there's a lot things we can control in this life, and one of them is who we choose to be. That reminds me of a story, about some friends of mine who never stopped believing they could make a difference. They went to school right over there . . .

HE POINTS AT SCHOOL SIGN.

                      BOY
        Cuckooville Elementary School!

THIS WAS *one* face of Neddie Marks, the second-grade movie narrator, the one with the message about the choices we have and the difference we can make in our worlds.

There was a very different face of Neddie Marks in real life, however. Neddie was about a foot taller than all the other second-graders. He was the one teachers shook their heads about. "I can see him being in prison before he finishes high school . . . if he lives that long." Neddie, it was said, had some problems at home and was a continual challenge in classroom settings—disruptive, attention seeking, and often intimidating to his classmates.

Although Neddie was a part of making the second-grade movie, he required special treatment—more often than not, placement in a different setting because of his disruptive behavior.

But you wouldn't dream of this "face" of Neddie if you had first seen his "movie face." The movie-writing process provided many interesting opportunities like this for the second-graders as they played with everything from their identities to their personal philosophies to their takes on popular culture to their understanding, or lack of, curricular concepts.

Conceiving the Second-Grade Movie

Diana Green already knew about the whole "moviemaking thing" before I shook her hand on the first day of class. She had seen the kindergarten and first-grade movies the teachers and I had made and was anxious to "continue the tradition." She was very comfortable with media technology, especially compared to her primary school colleagues, and was very excited about applying videomaking to her lesson plans.

But we didn't start videomaking immediately. We got to know each other in the context of her class. I worked myself into her class setting, from assisting students in small-group activities to reading stories. In all of the movies before this, getting to know the kids in the class environment seemed to quickly build a bond of trust. As a teacher myself, I could relate to the insecurity of a trespasser in my teaching environment. I felt it was important to gain the trust of the teacher because, without it, I would likely be seen as threat—an outsider who could potentially pass judgment on her teaching job.

It didn't take long to get comfortable with Ms. Green. We were already well into discussing ideas for the movie by my second class visit. From the start, she thought it would be great for us to do the video in the context of a national scholastic contest with the theme "Show how kids can make a difference in the world." She immediately saw connections we could make with her English Language Arts and Ecology curricula.

"We could shoot some video of them doing anything from picking up trash to helping out at community centers," she remarked on one occasion. I countered with, "How about doing a video to link our class with another second-grade class anywhere in the U.S.? It could be like a video pen-pal thing."

We continued brainstorming during class activities but if we were to enter the contest we would have to postmark the completed video by February 15. It was early January, and the children had just finished their winter break. We knew we had to get started on the video or it would never be finished on time.

On or about the week of January 9, one of the children in class mentioned that he had been thinking about an idea for a movie while he was working on a pollution poster for his class. "What if we did a movie about a bad person who was always polluting, and a bunch of kids dress up like a pollution monster and scare him so that he stops polluting?" This idea moved in a different direction than both of ours, more toward child fantasy. Ms. Green thought it sounded interesting, so much so that we decided to use it as a springboard to officially introduce the movie project to the class. A "monster-based" tale was certainly something second-graders could relate to and it felt appropriate to develop the story from the seed of a second-grader's imagination.

Involving all the students in creating the movie story would also fit in nicely with one of the class's culminating projects: writing individual storybooks. This video project would serve as an introduction to the basic elements of story structure that they were just now getting into. These elements included

- Setting
- Characters
- Problem
- Solution
- Ending

She saw the storybook project as a particularly effective part of her second-grade English Language Arts curriculum. Once the children wrote the stories, they would print them, create colored illustrations, and bind them. She often remarked how proud they were having written their own books.

In addition, such an examination of storytelling helped illuminate writing strategies and structures to the students that would improve not only their writing but also their reading. She called it the double-sided coin:

> Children who are good writers become better readers because they know
> how to look for information. It's like a double-sided coin and they both
> need to be there.

That's why she invested so much time on writing exercises in her second-grade classroom. She wanted them to identify with these concepts as both writers and readers so they all had to read each other's stories. Writing their own stories would provide an individual experience in storytelling. The movie, she hoped, might provide an additional storytelling experience they could share "out in the open" as a class.

A Different Kind of Writing

The writing process used in producing moving-image media is both similar to and different from writing for print media. There are similarities in the structures of both discourses—for instance, in Diana's storytelling terminology. Settings, characters, problems, solutions, and endings are as important considerations in movies as they are in storybooks.

David Gauntlett didn't expect the content qualities of his moviemaking experiences with English schoolchildren to outweigh the technological qualities, but they did.

> The production of environmental videos was a process which, perhaps surprisingly, was not centrally focused on technique and the technicalities of production and would be better characterised as discovery—albeit about the world, rather than the self. In this, it would seem to differ from school written work, which Pam Gilbets's survey of research suggests—in secondary schools at least—is far from being the personal, liberating experience celebrated by English educationalists: "Studies which interview students overwhelmingly indicate that students consider school writing tasks uninteresting and teacher-directed. . . . Students connect school writing with spelling, grammar, neatness, set length essays, and 'being correct'—not with the general liberating concept of writing as self-expression espoused in popular practitioner guides." (1996)

The difference resides in the respective crafts and technologies surrounding each writing instrument. Even though printed words are used in movies and storybooks, they serve more as a blueprint for the movie than as an end product. The end product is a conglomeration of pictures and sounds that evolve out of the words.

There was a time not long ago when this process and its accompanying technology was restricted to professional settings, given its high cost and complicated technical production processes. When I began making movies with kids in schools—in the early 1990s—I had to bring special equipment into the classrooms and edit the footage using the university's postproduction facilities. At this point in time, television production was considered a technology and practice of the privileged.

Now, every school I work in, big budget or small, possesses the necessary technology to produce and edit basic video productions—much of it at no cost.

Over the span of my thirteen-plus years of work with K–12 settings, visual production tools have become exponentially more affordable and accessible. With the slow eradication of the technology barrier, the challenge has shifted to finding effective educational applications.

Ms. Green had taken care of this with her movie idea and its connection to her writing and ecology lesson plans. Now it was a matter of turning her students loose on the project and hoping for the best. Though comfortable with video production as a class activity, she was very unsure how to do it and worried whether it would be too complicated for her students to figure out.

I assured her that the best thing we could do was to channel the experience in line with her instructional objectives. This would make the technical process secondary, and the writing experience primary. We decided to get started writing the movie in three groups of seven or eight children.

Setting Them Loose

When it came to writing the movie story as a whole class, we employed brainstorming technique: giving the groups basic questions or problems and having them generate as many and as varied responses as they could, without immediately deciding which was most appropriate. All answers were given respect (no matter how outrageous) and then the group worked toward a decision by consensus. This approach generated both doable and not-so-doable ideas, and imagination was by no means in short supply.

We divided my hour-long time frame with the class into three twenty-minute brainstorming sessions, one for each of the groups. While one group met with me, the other students worked on other class assignments with Ms. Green and her assistant.

I began with the first group by addressing the big question, "What do you want to do a movie about?" Ms. Green had already introduced the very basics of the "pollution monster" idea and asked them to think of ideas of their own, but their first suggestions were in strong support of the monster idea. It was clear that these children were very excited about the idea of making a movie about some kind of monster. And they were good about keeping their ecology lesson in view of their imaginative suggestions.

"Our movie should help people to stop polluting the Earth," Neddie responded.

"We should have a clean-up, Super Pollution Man," added Donavan.

"Help Mother Nature! Clean up oil!" yelled some of the others.

The suggestions quickly refocused on the superhuman powers of a hero figure. "Our Super Pollution Man should have pollution powers," said Neddie.

"And there should be a Pollution Woman," added Erin.

"And they should have a Pollution Dog," said JR.

"And a Pollution Cat," added Erin.

The entire group was truly enjoying the process of building a story idea. Voices were rising. The once-seated group was now standing. All of the children had their hands raised, pleading to add more ideas to the pile. At this point, one of the teaching assistants from the adjacent class poked her head into the room and politely reminded us that we should be careful not to become much louder. Her class could hear everything we were doing.

As we assembled back to reasonable order, Neddie announced he had a great idea. "We should have a Pollution Singer who sings the pollution song. He can be a rapper! [Rapping himself, while dancing across the classroom floor] My name is Neddie! I'm full of spaghetti!"

Recognizing at this instant that Neddie had just come up with (accidentally or otherwise) an extremely useful and creative visual storytelling device that often eludes even a good storyteller, I—along with my accompanying noisemakers—mouthed-out a percussion beat for Neddie to continue his rap with.

Neddie's rap could serve as an expositional vehicle or an efficient and enjoyable means of getting necessary information of a story across without reading that information verbatim. Neddie had stumbled upon a major difference between storybook writing and movie writing.

Television and motion picture media are celebrated for their abilities to show a story rather than tell it. In order to do this, visual storytellers must disguise story information, such as

- Who are the characters?
- Where did they come from?
- What motivates them to act as they do?

The real challenge here is that this must be accomplished within the action of a story. If there is one thing that does not work in television and cinema storytelling, it is long spans of stationary "talking heads" or traditional book-reading storytellers. Neddie's rapper could function in the same role as a traditional storyteller (only much shorter—roughly "book-ending" the story at the very beginning and end) to fill in the expositional details of our story the action could not get across. He could be a character in the story, setting up our story as a witness account—someone to assure us that our story and its message were clear to a viewer.

By this time, our creative story conference was completely out of hand, and our twenty minutes were up, so the next story conference group was brought in. To "distract" the rest of the class from the creative meeting going on in our corner of the classroom, Ms. Green decided to have students sketch their visions of the movie's monster. They happily agreed.

The next group addressed the topic of "characters." Who or what did they want the characters of the story to be? Like the group preceding them, they quickly gravitated to the subject of "the monster." They wanted a monster, but not just any monster. Michael described it as an "Alien Toxic Pollution Monster!" They wanted this monster to be a bad monster, not a good monster as was suggested in the original story idea where the monster was actually a group of kids *disguised* as a monster to scare away a bad kid who always polluted. Michael's version was the epitome of a bad monster, representing the evil force behind pollution.

"What other characters should be in the movie?" I asked. I shared the previous group's idea of a "good guy," Super Pollution Man and Woman (deliberately leaving out the dog and cat idea, as I thought we had our hands full as it was just working with second-graders). I further suggested that we think of making a handful of characters, as opposed to just one or two, be our superheroes. That way, more children in the class could be characters in the movie, if they so wished. This brought on the inevitable response from Willie, "Yeah, we could be the Power Rangers!" Another frenzy began, but was quickly brought to order as Ms. Green and the rest of class were now sharing the same space with us. The *Power Rangers*[1] happened to be one of the most popular television shows, particularly for elementary-aged children, at the time.

"So who are these Rangers? Where do they come from? Are they second-graders?" I asked.

A resounding "Yes!" streamed out from the group, followed by assorted claims on particular Ranger colors: "I want to be the red one!" "I'm blue!" "I'm pink!" "I'm white!"

"Don't you think our Rangers should be different from the Power Rangers?" I suggested. This both perplexed and disappointed them.

Reading Meets Writing: Conversing with Popular Culture

This was one moment among many where students were looking to popular culture works for ideas to apply to their story. This brings up an interesting question that Pat Kipping brought up in a media literacy listserve conversation:

> When given the opportunity to represent themselves in video, do children ever get beyond imitating what they see on TV and re-presenting themselves as portrayed by adults? What can one do to move through and past that stage, which in my experience seems unavoidable? (1995b)

In the case of the Power Rangers, the children were certainly headed toward imitation. This could be looked at in one of two ways: first, as a void in creativity on the children's part. Perhaps they were so saturated with popular culture like the Power Rangers that they had lost the ability to come up with anything original with their own imaginations.

I might be inclined toward this conclusion if I had not witnessed intelligent and creative college students and even TV and filmmaking professionals doing the very same thing when they faced a creative challenge. The first test of creativity almost always involves confronting or avoiding imitation.

A second way to look at "the imitation challenge" is to consider it a point of engagement in writing and reading processes. Why wouldn't young videomakers imitate if popular media is all they really know in the way of story content? Television and filmmaking have long been "read-only" media and we are just beginning to see the opportunities for children to participate in the "writing" side of these media forms with the growing ease and affordability of video-production technologies. Given this, a good degree of imitation should be expected simply because this kind of writing is so new to them.

But another aspect of dealing with this reading/writing engagement is the opportunity it presents to share in the negotiation of cultural meaning. Talking to them about the media they watch can help teachers and adults understand how children think of the world around them. This puts teachers in the position of getting to know them better, talking to them about issues and ideas in their worlds, and helping them develop their own perspectives.

In this light, the question of imitation is not simply to eliminate it. Imitation presents an opportunity to get into the minds of young readers and writers and illuminate their individual viewpoints and needs. This is not to say that getting beyond imitation is unnecessary—it is simply not the only beneficial outcome. Danielle Ippolito's sixth-grade autobiography project was a good example of this.

All about Me

This was an idea Ms. Ippolito came up with to blend her Art and English Language curricula. The students named the project *All about Me*, and the object of it was to get them to express something about themselves to the public—something that communicated some essence of who they were. The final work would be a collection of one- to three-minute autobiographies of everyone in the class.

Although this was very different from the Pollution Rangers case, the temptation for imitation—even with something as expressively original as an autobiography—was very similar. One boy, Joshua Teall, came up with a storyboard for his autobiography segment that involved martial arts moves, punching, kicking, and general violence between him (the protagonist of his story) and his enemies on the playground that sought to destroy him.

When we read his idea out loud, I could sense Ms. Ippolito's quiet discomfort with the violence. In one sense, it was a pure imitation of just about every cartoon in existence. In another sense, it was troubling that, at least on the surface, this boy saw violence as in some way an expression of who he was.

Based on experience, neither the teacher nor I panicked. We used the opportunity to engage Joshua and the class in a conversation about what he was attempting to communicate about himself. The program structure of *All about Me* was providing Ms. Ippolito the opportunity to do this without feeling invasive.

What we discovered in the conversation with Joshua was that the story was more of an expression of his love for drawing comics than a glorification of violence. He was not a violent guy, but he was proud of his drawing talent and hoped to one day become an accomplished comic artist or cartoon animator. This gave us much to consider as he shaped his visuals to best reflect his sense of himself. For instance, it would be important to include an interview segment with Joshua where he talked about his love of drawing cartoons, in addition to the cartoons themselves.

Creating the Pollution Rangers

Now that the idea of the Power Rangers had been thrown out to the group, the second-graders couldn't imagine anything else.

"How about a different kind of ranger? Instead of Super Pollution Man how about something like the *Pollution* Rangers?" Many of them smiled. They liked the ring of that, but they didn't take it any further.

"If there were Pollution Rangers—second-grade Pollution Rangers—where would they come from?" I asked.

"Cuckooville Elementary!" exclaimed Michael. The rest of the group laughed out loud. "Are they in the same class?" I asked.

"Yeah, Ms. Green's class," replied Shanara.

"Well, if they go to a pretend school, shouldn't the teacher's name be pretend too?" I suggested.

"Ms. Drazzle," responded Michael.

"Ms. Lanzer," added Emily.

"Ms. Jewls," added Meredith.

"These all sound like very good names," I remarked amazed at their ability to make up names.

"Too bad they're all characters from stories we've been reading!" explained Ms. Green from the other side of the classroom. "I know you can do better than that, boys and girls!"

"I know!" exclaimed Cory. "How about Ms. Book?"

"Or Ms. Encyclopedia," suggested Emily.

"Or Ms. Brook?" added Shanara.

"How about Ms. GooGoo-GaaGaa!" concluded Cory.

They all laughed, and it was time to bring in the next writing group. After being brought up to speed on story developments thus far, this group addressed the question of the central problem of the story. The problem would drive the action of the story and make for interesting interaction between the characters. Hopefully, it would also give us the opportunity to create a solution in line with our movie's overall objective: to demonstrate how kids can make a difference in the world.

Hannah had an idea right away, "Everyone throws trash everywhere!" After engineering a clever motive behind trash throwing, they really felt the need to address the design of the monster. Their deliberation over the monster pictures they had drawn took the rest of our time.

Structuring Imagination

Before we even turned a camera on, Ms. Green and I met to make sense of the ideas the children had come up with. We also needed to make sure curricular objectives were being met. This was where two major forces behind the project converged: the agency of a child's imagination and the structure of an adult's objectives.

The challenge was to preserve both in the moviemaking experience. As far as Ms. Green was concerned, and I wholly agreed, the project needed both. This challenge was becoming a universal phenomenon of K–12 videomaking.

Ms. Green and I tried to preserve the children's imaginative ideas while at the same time connecting the videomaking with her lesson plans. I weighed in on the practical side of determining how doable or undoable an idea might be in terms of video production.

On the Super Pollution Man and Mother Nature front, Ms. Green felt we needed to share the spotlight more—say, seven or so *star* characters instead of one or two. She immediately recognized that characters could be used to affect the classroom experience if we carefully cast the students. For instance, she saw the opportunity to put girls in leadership roles. She thought this was healthy and appropriate for this particular age group and their social development. She also wanted to exploit the opportunity to reverse roles of students: for instance, putting students who were known as "followers" in character roles that portrayed them as leaders, and vice versa. She felt it was important to use the experience to challenge everyday assumptions and labels that often were misleading, and even unfair. In a later interview, I asked her to elaborate on her motives behind casting the movie:

> I asked who didn't want to do certain things, and then I think I chose it with a multitude of criteria. I wanted to look multicultural. I had my own particular agenda. I said early on that I wanted a girl to be the hero. And that had some effect on it. I wanted to give kids a chance to shine, who maybe weren't good readers or good mathematicians, but could do a part in the movie.

In addition to casting, there were other points of conflict between child and adult perspectives. The children truly delighted in the action orientation of the *Power Rangers* show they were loosely basing their movie on. Generally, when the Power Rangers faced a problem, they approached it in an action-oriented—some would argue "violent"—manner. There were plenty of fighting sequences that sometimes made it seem these characters could kick and punch their problems away.

This was something about which we could offer competing notions and reinforce the idea that violence was not the only way to solve problems. An alternative solution to get the students to contemplate was, "Let's think about how we can solve this problem before breaking out with Power Ranger moves. We can sit down and have a meeting, think about it, and solve our problems with our heads, not our fists." This helped the children think critically about the notion of problem solving in storytelling.

Diana took pains to make sure the female characters were perceived as thinking on their own and in leadership roles. She was using casting to question dominant tendencies in both popular media and everyday life.

On the practical side, it didn't take long to dismiss the animal character suggestions, but we could probably pull off a monster of some sort, especially with a reasonably high "suspension of disbelief." We could strap two kids together in one costume as "the two-headed, alien toxic pollution monster," and it would not be difficult for a viewer to see that the monster was two second-graders bunched up into a one sweaty costume. Viewers could be expected to suspend the obviousness of the monster ruse, and still enjoy the story for what it was—a playful glimpse of second-grade imagination. In a case like this, the children's imaginations outweighed the need for adult structure.

Finishing Touches on the Story

After the meeting with Ms. Green, there still were unresolved story elements that we brought back to the creative teams.

The Pollution Rangers would essentially be a "special club" of girls and boys in Ms. Green's second-grade class who all shared a deep commitment to a clean Earth. The follow-up creative group decided to distinguish the Rangers not by function or special powers, but rather by name. There would be seven Pollution Rangers, named after the famed seven seas. This linked the storytelling exercise with their Ecology studies unit. They actually couldn't name all the seas so they had to visit the library and look them up. Even when they looked up the seas, they reported, there were more than seven, so they chose their favorite seven: Atlantic, Pacific, North, Mediterranean, Indian, Arctic, and Antarctic.

Over the course of the next week in ten- to twenty-minute sessions, we fleshed out an outline of the movie: *Pollution Rangers of the Seven Seas.* The "moral of the story" was a given—the contest theme, to show how kids could make a difference. The plot and characters were heavily influenced by popular culture of the time. The Ecology unit also played an important role in the story, thus the "pollution slant" to the story.

- *Setting:* The story took place in an elementary school, like theirs, but not quite theirs, so they created "Cuckooville" Elementary School. They had been heavily influenced by the sarcastic schoolbook series they were reading—a Ms. Green favorite—the *Wayside School* series by Louis Sachar.
- *Main Characters:* The main characters were the Pollution Rangers, loosely based on the high school–aged Power Ranger characters, only these rangers were much younger and *without* superpowers. Instead of general responsibility for saving the Earth, a more focused mission in "keeping the Earth clean" was the mantra of this team of heroes.
- *Other Characters:* The Rangers would need a teacher, park ranger, and, of course, some kind of evil villain-monster figure. In this case, it was going to be a "Two-headed, Alien, Toxic Monster."
- *Plot:* One day, a two-headed alien toxic monster lands his spaceship on Earth to set the groundwork for an evil plan: to fill the Earth with all the garbage from Planet Junko. To do this, he has to recruit Earth children and turn them into zombies that will pollute on command. Their zombie drone is a monotone "Must Pollute." A special group of students in Ms. Green's class realize that they have to do something to stop this evil plan from unfolding so they get together and try to make a difference in this threat to world ecology.
- *Problem:* Will the Rangers be successful? Will they find a way to make a difference? This is the main problem of the story. The Rangers first decide to try out some Power Ranger–style solutions to the problem, using force to stop the zombie recruits from polluting. When this plan falls on its face, the students must reach deep within their capabilities to creatively resolve the crisis.
- *Solution:* They realize (with the help of a tiny seer in a bottle) that clean, pure water is the answer to their problem. If they spray water on the zombie children, it will revive them to their original state and save the Earth from sure destruction.
- *Ending:* After transforming the zombies, the Rangers turn their attention to the two-headed monster and, after cleaning it up with some pure water, befriend it and send it back to its planet with some clean and pure water of its own.

Authorship

In making movies in K–12 classes, my role has evolved into a negotiator of sorts between three forces: student (storyteller), teacher (learning authority), and media (expressive tool). The question often arose as to what the right ratio was, or what the proper blend of forces should be in the final product. The school context made it clear: the lesson plan had to be the priority. The expressionistic opportunity for children's voices and the opportunity for them to learn media-production concepts, though important, had to remain secondary to the lesson plan.

Overall, what I thought might be a negative idea of an adult suppression of fresh, honest, untainted children's ideas turned very positive. Conversing with students and negotiating with them through the writing process allowed for what might be termed "critical alternatives," or discussions of incomplete notions or problematic outlooks children might have of their worlds.

Structuring imagination allowed for competition and confrontation with popular conceptions of the world. So it was not necessarily bad to enforce adult structures over children's ideas, but rather good to enlarge critical outlooks of not only media but also everyday ideas.

The K–12 Aesthetic

It was a Herculean effort by all, but we finished the movie, met the contest deadline, and premiered it for the whole school. *The Pollution Rangers* was a resounding success as a movie, and Diana and I learned a great deal through the experience.

In the end, we learned that children are more than capable screenwriters. Diana was extremely impressed at how uninhibited they were, especially compared to her, in putting their script together. They seemed to have a natural flair for thinking in pictures and sounds.

They also were acutely aware of visual storytelling processes based on their viewing experiences. They were actively engaged in the media they watched and applying it as they made sense of their worlds. They were fascinated with the opportunity to write their own movie scripts, in part due to the novelty of it, and also the opportunity it provided them to express their often ignored views on the world.

In the end, the writing experience was at the center of their fulfillment in the movie project. It was their movie, created by them—certainly with adult contributions and restrictions—but something that they took ownership of and felt proud of authoring.

It was through this project that I began to realize in my research that the aesthetic measure of K–12 media production was decidedly different from those I had worked with in the past.

Over the years of producing video projects with kids, my production mindset had shifted from mainstream entertainment technique to organic technique growing from the unique mind-sets of children and their learning environments. This organic orientation had proven very effective in K–12 settings.

As I have found over time, there does not have to be an elaborate screenplay to have a great movie experience. For instance, taping and watching (and rewatching) footage of a class engaged in a field trip to the zoo is aesthetically pleasing in this K–12 world.

There is an interesting realization in how easily pleased students are as long as it is their life, their shared experience, or their ideas that are mediated. Formal aesthetics don't matter as much in this context. The "scale of beauty" in a K–12 video is far more forgiving in educational settings than in professional storytelling settings. The final product is driven by their ideas and adventurous spirit, rather than by "how good it is" in an "on-air" sense. There is no such thing as "bad" video made by kids.

I thought about the *Composing Project* a local musician had arranged with similar-aged kids that year. She put the kids in the position of creating their own music, acting as facilitator, but at the same time opening up their sense of musical authorship and preserving it for others to hear. It was uplifting and beautiful. I realized I had facilitated in a similar way with *The Pollution Rangers* as I helped guide students and teachers through the technical gamut of the production process. In both cases, the most important aspect of the experience was the children's authorship.

Writing as a Different Face

Ms. Green and I realized we were seeing a "different face" of the children in much the same way as we had seen Neddie Mark's metamorphosis from problem child to "movie star."

Neddie's schoolmates certainly noticed, according to Diana. They stopped Neddie, who doesn't often receive compliments, and told him they loved his movie.

One schoolmate remarked that Neddie's performance surprised him because he always goofs around, and to see him doing something that serious was "really cool."

A parent of one of Neddie's classmates remarked, "I was pretty pleased with him, I mean he did a good job . . . he's a wild man, and for him to take that role I thought was pretty good!"

Even Principal McManus had something to say about Neddie: "I think the kids here, seeing their peers in the movie was important. It was a nice thing for Neddie in the movie. He's certainly had difficulties here over the years, but the accolades he was getting here from his peers was just such a great thing for him."

```
AS THE NOW VERY CLEAN TWO-HEADED ALIEN TOXIC POLLU
TION MONSTER BIDS THE RANGERS A WARM GOOD-BYE ON THE
PLAYGROUND, THE NARRATOR (NEDDIE) POPS HIS HEAD UP
INTO THE FRAME.
```

                            NARRATOR
     Isn't that a nice story? You see, we can make a dif-
     ference if we just put our minds to it. It may not
       be easy, but in the end, love will save the day.

NARRATOR FROLICS THROUGH THE SNOWY FIELD AS WE FADE TO
END CREDITS.

                            THE END

Note

1. A live-action children's entertainment series featuring a group of young teenagers with superpowers and a mission to defend all that was good from the forces of evil.

CHAPTER 6

BONDING WITH CAMERAS

> The important thing is not so much that every child should be taught, as that every child should be given the wish to learn.
>
> —John Lubbock

HAVING BEEN through it once myself and three times as a parent, I think I can say I understand what first grade is all about. It has a lot to do with the beginning of "real" school, where school crosses the line from play to work.

But along with this newfound seriousness comes a warm assurance that fewer toys doesn't have to be a bad thing. The first-grade teachers that I observed were keenly sensitive to the fragile mind-sets of their young learners. First grade to them was a crucial defining point for the idea of school and the learning process, where their young students forged notions of what learning was all about. Their job, as they saw it, was to nurture the bond between learner and school, hopefully to plant the seeds of lifetime learning.

This process of defining *school* was one of the most striking aspects of my experiences with first-grade teacher Sally Beechwood. It was Mrs. Beechwood's first-grade story that had reached mythical status by the time we finished our third movie together. Mrs. Beechwood was the legendary movie character who jumped out of her classroom window. And mythical status is a special achievement indeed when it comes to impressionable notions of what real school is.

Chapter 6

<u>IT IS THE LAST DAY OF SCHOOL AND MRS. BEECHWOOD IS READY TO DELIVER THE "LAST TALK" TO HER STUDENTS AS FIRST-GRADERS.</u>

THE NEXT TIME SHE SEES THEM THEY WILL BE BIG SECOND-GRADERS, AND THIS MAKES HER VERY MELANCHOLY. SHE IS GOING TO MISS THIS CLASS. SHE CALLS THEM "HER FAVORITES." MRS. BEECHWOOD TAKES A DEEP BREATH AND BEGINS TALKING.

> MRS. BEECHWOOD
> "You've all been such good first-graders this year. I'm really, really going to miss you. And when you're all big second-graders I hope you'll come visit me and tell me what you're doing. I want you all to keep writing in your journals and have a great, great summer."

ONE BY ONE, THE STUDENTS FILE OUT OF THE CLASSROOM, AND MRS. BEECHWOOD WISHES THEM EACH A LOVING GOOD-BYE WHILE THE RAFFI SONG *"SO LONG"* PLAYS IN THE BACKGROUND.

AFTER THE LAST STUDENT LEAVES, MRS. BEECHWOOD WALKS TO THE BLACKBOARD AND ERASES THE CHEERY MESSAGES:

> *Goodbye to my favorite class!*
> *Have a great summer!*

SHE CONTINUES SADLY OVER TO HER DESK, SITS DOWN, AND LOOKS OUT AT THE EMPTY DESKS IN FRONT OF HER, THE VERY DESKS THAT HAD BEEN FILLED WITH SWEET, SWEET FIRST-GRADERS ALL YEAR LONG. SHE IMAGINES SHE CAN SEE AND HEAR THEM ALL, SMILING SWEETLY, WAVING *THE GOODBYE SONG:*

> Goodbye, goodbye,
> have a happy afternoon.
> Goodbye, goodbye,
> I hope to see you soon.
> Goodbye, goodbye,
> go safely on your way.
> Goodbye, goodbye,
> I'll see you another day.

SHE SNAPS OUT OF HER REVERIE. THERE IS SOMETHING OUT-
SIDE HER WINDOW. DID SOMETHING JUST MOVE? DID SHE
IMAGINE THAT NOISE? SOMETHING WAS SURELY GOING ON
THERE, OUTSIDE HER WINDOW. COULD IT BE . . . ?

YES, IT WAS! FIRST ONE HEAD POPS UP. THEN A SECOND,
AND THIRD, FOURTH, FIFTH . . . AND THEN AS THE FINAL
FIRST-GRADER'S HEAD POPS INTO VIEW, THE WHOLE GROUP
ERUPTS IN A LOUD CHEER. MRS. BEECHWOOD RACES TO THE
WINDOW.

                    MRS. BEECHWOOD
                    You're back!

HER "FAVORITE" FIRST GRADERS WERE OUTSIDE BECKONING
HER TO JOIN THEM. WITHOUT HESITATION, MRS. BEECHWOOD
JUMPS OUT OF THE WINDOW TO JOIN THEM. CROWDING AROUND
HER, ARM IN ARM, MRS. BEECHWOOD AND THE CHILDREN WALK
UP THE HILL BEHIND THE SCHOOL INTO THE HAPPY SUNSET OF
FIRST GRADE.

Bonding with Cameras

Bonding among students and their teachers was a universal phenomenon of K–12 videomaking, regardless of grade level. There was something about the shared experience of making a movie and then watching it together that seemed to generate across-the-board positive feelings in participants.

Part of the bonding experience for K–12 students involved an appreciation for the people they shared the experience with. The children felt they had shared an unforgettable experience, and because the experience was on videotape, it could easily be watched again over the years, ensuring its placement in long-term memory. There tended to be two types of bonding: bonding through the work of videomaking and bonding through viewing.

Bonding through Videomaking

> It was different than other years. The movie was a lot of work, a lot more work than the children foresaw. . . . It was also a wonderful bonding experience for them and the class really seemed to come together as a whole, better than other years.
> —Ms. Green, on the making of her second-grade movie

Mrs. Beechwood's kids and I spent a lot of time exploring the effects of various video-production techniques, from lighting to composition to motion to sound. There was a certain degree of visible pride in their slow but sure discovery of the secrets of the trade. This was evident when they excitedly revealed to their family and friends how their videos were produced.

After I had worked myself comfortably into Mrs. Beechwood's classroom routine, we talked to the children about making a movie. The children's first idea was inspired by the popular mystery book series *Goosebumps*, by R. L. Stine. This series was at the peak of its popularity, and very fresh on the first-graders' minds.

They wanted to call it *Beechbumps* in honor of their teacher and they wanted it to be about everything. At first, they liked the idea of a hosted show starring Mrs. Beechwood, but after talking about it a little they realized it would be better if *they* were the stars. They brainstormed a rather eclectic list of ideas for the show:

- Magic
- A song
- A giant story
- Movies, movies, movies (like the popular ones they knew at the time—*Mortal Kombat, Star Wars, Ace Ventura*)
- A *Goosebumps*-type story
- Comedy acts
- An interview
- News
- A cooking show
- Something about animals

Almost all their ideas were influenced by popular culture and fantasy, so the first thing we did was to talk about the programs they watched and how they were made. The idea was to help them develop a critical perspective on television, to demystify it at the very least.

We broke out the camera and demonstrated how video stories were produced, particularly how TV filmmakers used the camera to make kids feel particular ways. For instance, we talked about the use of *angle* in representing a subject on camera. If a producer wanted to make a person appear powerful, he or she placed the camera below the subject's eyes, and vice versa if he or she wanted to make a character appear more vulnerable. We demonstrated this on the biggest and smallest kids in the class by reversing impressions of them, making the smallest appear powerful and the biggest appear weak.

Although the kids seemed to get the point of the lesson, they were much more interested in simply seeing themselves on camera. They were having fun, but I worried it was at the expense of the lesson behind the visual activity.

Our next subject was *light*. I demonstrated how a filmmaker could use light to make a subject appear either soft and friendly or hard and unapproachable. We lit one of the kids with a soft spray of light that made her appear angelic, and another with a hard light from the floor that cast eerie shadows on his face and made him appear troubled.

We continued with more lessons in sound and motion. We demonstrated the effect of music and natural sounds on pictures. We played different musical pieces over the same video footage and talked about how the musical selections prompted different interpretations of the picture—happy music imparted happy feelings on the image, whereas sad music brought out sadness in the same image.

We showed how motion picture images were actually collections of still pictures presented rapid fire on a screen, and how we could utilize this persistence of vision phenomena in making a stop-action movie.

The university's TV studio was our next stop as we surrendered all the controls—cameras, lights, microphones, studio, and control room—to the wildly excited first-graders. It was at first, as it usually is with children, moderately to severely chaotic, but when the kids settled into their positions, we actually managed to make a little TV show. It was called *The Mrs. Beechwood Show* and it was hosted by none other than their teacher.

The show began with opening music and a title graphic over a shot of the set they had assembled. This was followed by an interview between Mrs. Beechwood and two of her students, and a magical flying segment. The flying segment involved placing one of the students in front of a blue screen and then "electronically keying" her over videotape footage of the sky, which made it appear as if she was flying.

By the end of the school year, our final list of show elements was surprisingly reminiscent of the original list, falling short on only a few of the items.

1. Stop-action movie with action figures
2. Scary tyrannosaurus story
3. "Monster behind the door" story
4. Slow-motion segment of kids playing and lining up at the end of a school day
5. Movie story about the last day of school
6. Music video with a collection of visual literacy shots: high angle, low angle, soft light, hard light, disappearing children, and assorted moving portraits of students in the class

In the end, when the first-graders looked at what they had done, they exhibited a deep sense of ownership. They were watching their stories and their ideas portrayed on screens normally reserved for professional works. They bragged about the movies they had made together to all who would listen, especially their parents. There was a clear sense of pride in their workmanship and this seemed to be a universal phenomenon in all grade levels.

The social bonding through workmanship was also something that seemed to surface in K–12 videomaking experiences other than my own. A former student of mine, Adam Stockman, became a public school librarian and created a library-based video club for students. He shared his perspectives on the bonding he saw through videomaking:

> Sure, kids are learning how to use the camera, but in the process they are learning how to work successfully with others, bring a complex project to fruition, and polishing their writing skills. It's exciting for me to watch kids get excited about a process rather than a result.

Across the Atlantic, David Gauntlett had discovered the same with kids in British school systems: "In addition to enlivening the meaning and relevance of environmentalism to their own worlds, the experience had a group effect: The enthusiasm of the whole group meant that all contributed something" (1996).

In the end, the experience of making a video was fun for the children, but it was also unexpectedly hard work for them. Perhaps it was the feeling of earning a reward for their work—a finished video—that accentuated their class experience and relationships with each other.

With Diana Green's second-grade movie, there was a line crossed between the excitement of the experience and the sheer work of it. Diana related this in a discussion about a particularly cold shooting day:

> The kids were so good about being outside with their coats off.... They clearly let us know that they were cold, but they were willing to put up with that.... They were troopers.

Another part of it was the continual self-discipline it demanded of them. Diana talked about the speech she gave to her second-graders:

> "This is what we're going to do. We're going to film this story and some of you will be in some scenes and some of you won't ... there will be a lot waiting time. And you just have to accept that. You'll be sitting there with friends, watching what is going on. It may happen fifteen times, but you still have to wait through it and be quiet. Because if you don't, you're going to ruin the movie and it won't ever get done."
>
> They were able to buy that. There were several instances it was almost killing them to sit still and watch through—and these are not kids with

very much self-control. They really were able to pull it off.... They were as perfect as they could be for second-graders.

One other aspect of the work the kids did was their "acting." They often had to role-play and pretend certain things that were simply not true. In this sense they were bonded by their shared conspiracy.

The Last Day of School was a good example of this. They had to pretend it was the last day of school even though it was not. The conspiracy seemed to provide them with power in the knowledge that they knew something a viewer wouldn't know, including how a segment was shot or the "white lie conspiracy" behind it. Even though it was fun for them to act, it was not easy. It took a good degree of restraint and self-discipline not to laugh.

In his Internet article "Meaningful Digital Video for Every Classroom," Hall Davidson commented on the hard work that goes with videomaking:

> In practice, videomaking in the classroom takes dedication, inspiration, and plenty of extra time, not to mention the additional management and equipment responsibilities it tacks on to your day-to-day duties. It is flat-out hard work. Despite those difficulties, teachers have always made great videos. Perhaps this is due to a combination of the tremendous appeal of video, the deep satisfaction of seeing stellar projects on the television set, and the knowledge that work can be archived in media collections. Plus, of course, the great educational benefits to students.... As with any videomaking, the process of creating the project is as valuable as the final project itself. (2004)

Over the years of working with teachers, I have spent much time and energy worrying about the extra work that goes with video production, looking for ways to simplify the process, especially for teachers. After all, teachers' jobs are hard enough without extra work. But thinking about Hall Davidson's and Adam Stockman's comments on the importance of the process—not just the final product—maybe the shared experience of hard work is one of the reasons why it works so well.

Bonding through Viewing

BEAUTIFUL BUNCH

```
                MRS. BEECHWOOD'S SECOND MOVIE
        A MUSIC VIDEO TO ACE OF BASE'S BEAUTIFUL LIFE
                 DIRECTED BY MRS. BEECHWOOD

FADE UP ON TWENTY-PLUS FIRST-GRADERS SURROUNDING MRS.
BEECHWOOD ON A TELEVISION STAGE AS THEY RECITE IN
UNISON:
```

                    Yeah . . . Beechwood Bunch!!

DISSOLVE TO A SHORT COLLAGE OF SHOTS OF MRS.
BEECHWOOD'S FIRST-GRADERS DOING THINGS TOGETHER OVER
THE MUSIC. AS THE SHOTS PROGRESS, A SERIES OF WORDS
OVER BLACK FORMS A QUESTION:

                            Who
                            are
                           these
                          people?

AN ANSWER FOLLOWS:

                    Why, it's the Beechwood Bunch
                           of course!

A MOTION PICTURE SLIDE SHOW OF EACH MEMBER OF THE
BEECHWOOD BUNCH FOLLOWS. THE FIRST THING WE SEE IS THE
BUNCH MEMBER'S NAME IN LARGE WHITE LETTERS OVER A
BLACK BACKGROUND AND THEN A SERIES OF SHOTS SHOWING
THE NAMED INDIVIDUAL INVOLVED IN VARIOUS ACTIVITIES
OVER THE SCHOOL YEAR.

WHEN THE SLIDE SHOW IS COMPLETE, THERE ARE MORE SHOTS
OF ALL THE KIDS TOGETHER AND THEN AS THE SONG FADES
OUT WE GO BACK TO THE TELEVISION STAGE WITH ALL THE
KIDS CROWDED AROUND MRS. BEECHWOOD AS SHE SAYS,
"LET'S DO IT AGAIN!"

FADE TO BLACK.

This video was based on the slide show Mrs. Beechwood had done in a past school year, *before* she met me. We talked all year long about experimenting with a video slide show that she could edit with nonlinear video-editing technology. It would be a slide show with a few extra features: moving images, music, motion control (slow motion, fast motion), and graphics. The motive behind the slide show was to instill a positive image of school experience in her first-graders. Mrs. Beechwood explained the slide-show project she had done in the past:

> I just took slides throughout the year. In fact, it had great potential, but first of all I'm not the best camera person and second of all, I would kind

> of forget. It would be like, "Oh—here's a great moment! I forgot the camera today." Or, "Here's a great moment! Oh, it's too dark."
>
> But I was smart enough to start at the beginning of the year and right through the end of the year. And you could see the progression and with children you could see how they grew—physically grew. It was a great reflection on the year.... And I did music—I just turned on the music in the background. It wasn't great, but it was better than nothing.
>
> They just were delighted to see each other and they seemed to take pleasure in seeing their friends. "Oh, look at that face! That's you!" and "There I am!" and they were a little embarrassed, because I took a lot of silly faces and stuff like that ... perhaps it gave them importance, because they were on film and it was important what they were doing and we did experience this and it was part of school.

An entirely different form of bonding seemed to occur around *the viewing* of school videos, as opposed to the bonding between working videomakers. This form of bonding involved not only the videomakers and their teachers but also parents, friends, neighbors, new students watching former students' movies, and even people who had no connection at all to the videomakers or their school settings.

In this case, the bond revolved around the experience of sharing a story. The most natural place for this kind of bonding to occur was during the school screenings of the videos. These ranged from small-group viewings to schoolwide ceremonies.

Corrigan Elementary Principal Carolyn McManus talked about some of the screenings that stuck in her mind:

> For me, the reaction from staff and kids has just been great. I think it's been a real big morale booster. The staff was just howling, to see the kids, to see Sally Beechwood ... I think people felt really proud about that, proud about the kids and felt really good about the movie.... I had taken the *Pollution Rangers* home and my kids watched it and they wanted to watch it over and over and over—they were fascinated by it, absolutely fascinated.

As Principal McManus indicated, the school screenings of K–12 movies were not the only bonding-through-viewing zones. For instance, Ed and Karen, father and mother of one of the children in Mrs. Beechwood's movie, introduced themselves to me while we were waiting for our respective children outside a classroom.

"We're David's parents," Karen led. "Thank you so much for everything you did on that first-grade movie. It's so great. You should see, the first thing we do when anyone visits our home is to sit them down and pop the movie in the VCR," she laughed.

I laughed along, partly because that was exactly what we did in my house when people visited us. I also realized that on a symbolic level we were sharing the same story, but deploying it in different, yet universally similar, ways. To me, my son's first-grade movie was an emotionally stirring account of *his* first-grade experience.

To Karen and Ed, it had become something of the same except with their son David at the center of the account—the selected protagonist. Yet, the purpose behind our simultaneous claims on the story could probably be termed naturally paternal. We were finding significance in the existence of our children and ultimately ourselves as origins of those children. In that moment, we were celebrating our respective stories and feeling some sense of human significance through them.

This was not my first realization of the emotionally charged bonding of videomaking in the school environment. The most moving example of this was the experience I had with Patricia Thompson. She was the mother of Rachel, the "star" of a kindergarten movie, and the unfortunate victim of a terrible tragedy a few weeks after the making of the movie.

While Patricia, her husband, Kenneth, Rachel (age 6), and Rachel's sister (age 7) had been camping, an unexpectedly ferocious windstorm swept a large tree onto their tent, instantly killing Kenneth and seriously injuring Patricia. Miraculously, the girls were not hurt.

All along the walk to Patricia's hospital room, I searched for the right thing to say, but found myself face-to-face with her as ineloquent as I had started. It turns out I did not have to think of much to say, because she took immediate control over the conversation. She introduced me to her sister as the man who had made the movie with Rachel. She explained that she had requested that her houseful of family and friends (awaiting her husband's funeral) watch Rachel's movie. "It's so wonderful, I want everyone to see it!"

I wondered how someone who had been through such a tragic set of circumstances could find any value or comfort in a comparatively unimportant video at a time of terrible loss. I was both mystified and moved.

I watched the movie again that night and saw it in a completely different light. I found myself focusing on the space/time contradiction between the painful present and the unknowingly sweet past. I wanted to crawl through the time warp between the Rachel I saw in the movie and the one I had seen in the hall outside her mother's hospital room. If only I could have warned her of what was ahead.

That night Rachel's aunt telephoned me with a request from Rachel and her mom. Since a number of family members would be unable to attend, Patricia thought it would be a good idea if the funeral ceremony could be videotaped for them to see at a later date. Rachel insisted that I shoot it.

Over the years we continued to share warm memories of the school movie we made, but we haven't talked much about the funeral tape. Obviously they're two very different memories representing two very different phases in Rachel's life, the latter (for better or worse) probably not the most striking conversation piece. However, Patricia did confide in me that her daughter was particularly moved by the fact that the cameraman cried at the funeral.

Such social bonds spurred by the K–12 movies seemed to demonstrate the power of the storytelling we were doing. The stories we were making were having impacts well beyond the small classrooms they were made in.

This was unanticipated on my part in the whole videomaking project. Once again, I had expected my work with the kids would be more connected to demystifying media processes so that they could watch television with informed, critical perspectives. My underlying theory behind this was that they would carry a healthy skepticism of media messages and ultimately demand more of content. If such a movement caught on, it could generate a new class of informed, young viewers that TV programmers would have to raise programming standards for.

But my work with the kids was going in a far different direction than I had anticipated. It was becoming a content-centered experience, heightening children's awareness of and involvement in the practice of storytelling. I realized this one day when I was in the middle of a storytelling lecture with my college students.

One of the main challenges in my college production classes is to get students to respect the power of storytelling. In general, they tend to see the process of storytelling as a fairly superficial process between a viewer's need to "escape" and the filmmaker's desire to share his or her creative stories.

What they don't tend to realize is the immensely important social process behind entertainment—it's not as superficial as they think. Viewers search for more than mindless entertainment. Simultaneously, they are engaged in a complex process of making sense of their lives and they look to stories to assist them in this task.

The definitions of storytelling I put in front of my college students were as relevant to K–12 production settings as they were to future professionals. I explain to my students that storytelling is not just entertainment. In *Tales of the Field*, John VanMaanen explains storytelling as

> the primary way through which humans organize their experiences into temporally meaningful episodes.... It makes individuals, cultures, societies, and historical epochs comprehensible as wholes, it humanizes time, and it allows us to contemplate the effects of our actions and to alter the directions of our lives. (1988)

In *Video Communication*, David L. Smith explains storytelling as a universal mirror that shows us the "truth" about ourselves—who and why we are. When we look into this mirror, we see daily routine and mundane circumstance transformed into something profound.... Through story we can transcend the experience of daily living and know ourselves as more enduring than the little occurrences that mark our individual existences. (1991)

The K–12 videomaking experience was becoming more of a storytelling experience in VanMaanen's and Smith's senses than a media literacy experience. There was plenty of media literacy going on, but unexpectedly, there was even more energy and resonance in the stories students were producing—more than I had thought was possible. The videos these K–12 students were making were becoming part of a collective consciousness that began in classrooms and spread to lives outside school.

As the mother of one of the student's so eloquently expressed, her daughter's school movies hit square in the heart:

> I sit there and cry, every time I hear the song. Like when I play it over with the kids, I start to ... awww [mimics getting choked up with tears] and think of the movie ... the slow motion with the kids and the music overlaid, and classroom shots. I think that was more what hit the mother in me.

It wasn't just the story of her daughter's first-grade experience; it was as much the story of her daughter.

Intangible Bonding through Videomaking

In our third movie together, Mrs. Beechwood thought it might be fun to do a sequel of the famous *Last Day of School* story that involved her jumping out her classroom window. It was six years after the first jump and the myth had been passed down through every class she had had since then. She had screened *Last Day of School* every year since its making, and the students always loved it. It was, after all, part of the Beechwood experience.

THE LAST DAY OF SCHOOL II

MRS. BEECHWOOD'S THIRD MOVIE
A MUSIC VIDEO TO VITAMIN C'S *GRADUATION SONG*.

OPENING GRAPHIC (WHITE LETTERS OVER BLACK BACKGROUND)

               THE LAST DAY OF SCHOOL
              MRS. BEECHWOOD'S FIRST GRADE

<u>DISSOLVE TO LONG-SHOT BEHIND A CLASSROOM OF FIRST-
GRADERS AS THEY LOOK AT THEIR TEACHER AT THE FRONT OF
THE CLASSROOM. MRS. BEECHWOOD IS GIVING THEM THE
FAMOUS "SECOND-GRADE TALK."</u>

                    MRS. BEECHWOOD
         Boys and girls, it's the very last day of
         first grade. . . . I'm going to be so sad. I'm
         going to be missing you all summer long. No
         smiling faces to come through my door. But I'm
         happy because you're so smart. You're all going
         to second grade and you know what that means. . . .
           On the last day of school, when you're
              getting ready for second grade. . . .
                    What's it mean?

               THE CLASS CRIES OUT IN UNISON:
                 We're going out the window!!!

THE CHILDREN WALK UP TO THE WINDOW AND JUMP OUT, ONE
BY ONE. AS THEY ARE JUMPING, WE DISSOLVE TO SHOTS OF
VARIOUS ACTIVITIES THEY HAVE BEEN PART OF OVER THE
SCHOOL YEAR, INCLUDING THEIR "VERY MOST FAVORITE"
ACTIVITY—GOING DOWN THE SLIDE.

BETWEEN SHOTS OF ACTIVITIES WE SEE EACH OF THE CHIL-
DREN SLIDING DOWN THE SLIDE IN HIS OR HER OWN PARTIC-
ULAR WAY—SOME SLOW, SOME FAST, SOME BACKWARDS, SOME
TWIRLING AROUND, BUT WE SEE EVERY ONE OF THE STUDENTS
BY THE END OF THE MOVIE.

IN BETWEEN THESE SHOTS, WE SEE SPECIAL CLOSE-UP
IMAGES OF EACH CHILD DURING THE SONG'S MAIN VERSE:

               As we go on we remember
              all the times we had together.
           As our lives change, come whatever
             we will still be friends forever.

THE LAST PERSON TO GO DOWN THE SLIDE IS, OF COURSE,
MRS. BEECHWOOD. ALL THE FIRST-GRADERS CROWD AROUND HER

AS SHE GETS TO THE BOTTOM OF THE SLIDE, AND AS THE MUSIC OF *THE GRADUATION SONG* FADES, SO DOES THE PICTURE.

THE END

This movie was scripted along the lines of the children's answer to the question, "What will you remember most about first grade when you get big?" Their answers included everything from routine class activities (such as playtimes, story reading, art projects) to outdoor recess to field trips. They even mentioned Mrs. Beechwood's famous jump from the class's ground-floor window, only this time, they wanted to be part of the jumping. That would make it *their* story.

In the end, the students were telling *their* unique stories, both as this year's "Beechwood Bunch" class and as individual members of the class, but they were also part of a larger story framed by their teacher—the story of *their* school. By making the movie and focusing on the warm and friendly icons of a typical school year, Mrs. Beechwood was creating an image of school for the students to identify with both in the present and in the future. The movie was the icing on the cake of what she hoped all her other activities were accomplishing—creating a positive image of lifelong learning and a positive impression of school experience. Hopefully, this was an impression that progressive grades would build upon like a foundation.

To characterize this experience as purely a social bonding experience might miss some of the intangible extras the bonding seemed to unearth. David Gauntlett found an example of this around his observations of bonding in his videomaking activities:

> On a broader scale, and most probably to a limited extent which should not be exaggerated, the video-making process may have contributed to the sense of community-feeling which children had for their area. . . . The generally positive view of each [school] area supplied by the videos may have led in some small sense, to a greater appreciation of the variety of people and faces who make up that community. (1996)

The bonding experience around videomaking seemed to open doors to kids' perceptions of and involvement in the world around them. Perhaps this was because in telling their stories they were realizing their significance and personal worth. They were seeing themselves as important parts of their worlds. They were authenticating their individual existences as storytellers and seeing connections between their authorship and the world of possibilities around them.

For instance, when they screened *The Last Day of School II* there was a class-wide gasp when the shot of Dougie, an autistic boy in the class who could not speak, appeared on screen. They seemed filled with warmth, almost at a parental level, that their special-needs classmate appeared uncommonly normal on screen. No one but they would know the everyday reality of Dougie—often challenging and disrupting to them. Dougie would be perceived by those who did not know him as "one of the bunch" and they delighted in this storytelling "conspiracy."

In this sense, their storytelling was autobiographical. They were telling their own stories, mapping out their existence, and seeing value in it. From what I could see, it appeared that the videomaking experience (both making and watching) was assisting the first-graders in actively constructing impressions of school. It was also strengthening their bonds with their classmates as well as their everyday morale, and helping them to understand themselves as part of a larger world. It was clearly strengthening their notion of school as a happy and safe icon of their lives.

Videomaking as Autobiography

In talking about the past, we lie with every breath we draw.

—William Maxwell (1980)

In their article "The Invention of Self: Autobiography and Its Forms," Bruner and Weisser (1991) use the familiar term "genre" to describe a means people use to formulate not only their stories, autobiography or otherwise, but also to organize their views of the world in general. Genres create possibilities, conventions, and expectation both in the construction and reception of autobiographical storytelling accounts.

In this sense, Mrs. Beechwood and her first-graders were employing the K–12 videomaking genre, along with its resident filmmaking and TV connotations, to spin their very own account of their first-grade experience. Granted, this was not a highly realistic or exactly true-to-life account, but it was true to the spirit of the children and their teacher.

Their autobiographical account was like any other, according to Bruner and Weisser—a form of "rhetorical strategy."

The point in this is that there is profound power in storytelling, in both the construction and recognition of stories.

One of the greatest values in Mrs. Beechwood's first-grade story was its near effortless repeatability. In video form, stories can be circulated and repeated time after time. This gives further strength to video accounts and spreads the bonds of story experience beyond its local origins.

Video preserves our stories beyond immediate and even long-term memory. As I write these words, I am reconsuming the stories of the first-grade videos and resupplying my memories of them with not only accuracy in account (which fades over time in my human memory) but also new experiences to juxtapose with these old stories—like the way that I saw Rachel differently after the tragedy in losing her father. This gives the video stories further strength in making sense of our lives, because their relevance extends beyond the time and place of their making.

In this sense, Mrs. Beechwood's famous jump out the window ages like wine, gaining value over time. I found it quite interesting that I had actually incorporated the jump into my own story over the years, in effect, making a little movie in my mind about it and filing it in my permanent memory. When I took out the actual movie to check on the details I *thought* I knew, I found that I had manufactured some things in the story, mostly emotional embellishments, that never truly existed. I had taken ownership of the movie in my mind and edited it to my own use in making sense of my world, my kids, and their school experiences, including memories of Mrs. Beechwood. I suppose it could be termed a home movie in my mind.

In my memory, I had polished up many of the original imperfections of the actual movie. I was surprised how many little mistakes in the actual movie my memory had corrected. I had even imagined a short scene or two that had never truly been part of the original movie. I was fusing the movie into a story of my life, in a similar way that the first-graders had fused their real first-grade experiences into a movie that symbolically represented their experience.

On the simplest of levels behind all this storytelling, videomaking participants were bonding together on a social level: student to student, teacher to student, parent to student, parent to teacher, and students, teacher, and parents to their school.

In the end, one of the most important features of this experience was its timelessness. The movies the children made are as meaningful to watch now, perhaps more so, as they were when they were created.

Mrs. Beechwood certainly has found this in the classes since our last movie together:

> This year when we get new students, I'll have probably four that come in and go "She jumped out that window once," because I show them the video . . . this year's class really liked it. [Laughing]
>
> Seeing me . . . they knew somebody in the movies!

CHAPTER 7

VIDEOMAKING AND CURRICULA

I hear and I forget.
I see and I remember.
I do and I understand.

—Confucius

FOURTH-GRADE teacher Erin Bronson was more intimidated by the technology factor of videomaking than any other teacher I had worked with before. She worried not only that she knew nothing about how to make a movie but also whether she could even determine an appropriate context for it in her class. She was, after all, a teacher—not a filmmaker.

What Ms. Bronson failed to realize, just like most of my introductory-level college students studying filmmaking, was that the true challenge of filmmaking had less to do with technology and more to do with content. In other words, the key to successful filmmaking has more to do with a filmmaker having "something to say" than in the filmmaker possessing technical filmmaking skills.

Erin Bronson, like most teachers I've worked with, had something to say. She had something to say because she had a clear mission as a teacher—to enlighten a group of children on various subjects, mostly along the lines of state-mandated curricula. Her fourth-grade curriculum delivered a wide range of learning objectives, or "things to say," from reasonably easy concepts to impossibly difficult ideas. Her job, like any teacher, was to bait her students into the "trap" of knowledge. She did this by devising lesson plans and class activities around her curricular objectives.

Based on past experiences (K–3), I encouraged Ms. Bronson to think of a lesson or curricular objective that video might enhance, perhaps even a topic that she found difficult or problematic in the past. The first thing that came to her mind was "research." Ms. Bronson thought that if video could help her students get more interested in research and study skills so they could see the value and relevance of these practices in their lives, it would be a great accomplishment. This was something fourth-grade kids were not typically enthusiastic about. Perhaps incorporating video into the research process would "charge up" this lesson plan for the students. Ms. Bronson reflected back on process of working videomaking into her curriculum:

> The kids were very excited. They just loved the whole idea and they asked right away, "Can we do it?" And it was just a matter of, well, I don't really have the skill to do the editing. So we started listing ideas of things that we could do and things we should include and . . . a format of how they wanted to do it. They wanted the idea of a documentary [so] that they would be teaching something and it would be something we could show kids year after year.

The class was working on a state history unit that focused on New York's Erie Canal. She wanted to use this unit to teach not only the topic but also specific research and study skills: how to access information and incorporate it into a learning activity. Students would receive a research topic and then go to the library to retrieve information about that topic, which would culminate in a written report. We determined that the difference with a video project would be that the report would be written for a TV news presentation. It would still be written on paper, but paper was no longer the ultimate destination for the words—the television screen was. As the project unfolded, Ms. Bronson also saw other curricular advantages:

> One thing that we need to work on is speaking skills. We need to be able to do presentation. The fact that students needed to be able to put together a project, the need to research . . . all these things were things we use all the time. . . . We have some kids that obviously speak and some kids who don't . . . but they were working towards that whole process and that was definitely part of our curriculum.

The video project was providing a playing field for the application of many more curricular activities than Ms. Bronson had expected.

The Relevance Factor

If there has been one finding more resonant than all others over my years of K–12 videomaking, it is something I call the *relevance factor.* The experience seems to make topics that video is associated with more relevant to young learners than they would have through more traditional instructional means. By the term *traditional means,* I'm referring to everything from teacher-to-student lecturing, use of written handouts, film strips, videos, audio recordings, field trips, and even the simple act of writing terms and concepts on a blackboard.

The juxtaposition of videomaking to curricular challenges and activities seems to consistently provoke an extra level of learning opportunity beyond traditional methods of instruction. This is where the idea of relevance kicks in. There is something about the way that ideas are presented to K–12 children that seems to connect with Confucius's wisdom on learning.

Based on observing so many levels of learning (K–12) over the past decade, it seems to me that learning is a process not unlike filmmaking. It is *not,* to say the least, a simple or straightforward process. Instead, learning is richly complex, multifaceted, and ambiguous.

There are no fail-safe steps to follow in order to make a great film. Some of the greatest known filmmakers will follow an acclaimed work with a work that critics call their absolute worst. If there were a way to make a successful film every time a filmmaker picked up a camera, they certainly would do it, but it just isn't that simple. Neither is learning.

When a filmmaker makes a film, she or he has to think like the brain of a viewer watching the film. In this sense, a viewer can be thought of as a disconnected perspective that must be lured into a story experience. To accomplish this, the filmmaker offers ideas of value and relevance to viewers. If it were as easy as stating those ideas upfront in concrete terms, most certainly would—but it is not that simple, and such is the case in learning as well. If it were as simple as placing the knowledge in front of the students so that they could absorb it and be done with the school day, teachers would do it.

Filmmakers and teachers must therefore apply their life experiences, abilities, and passions to engage their respective audiences in ideas. Both are in search of a complicated objective that is, borrowing from a branch of psychology,[1] a whole greater than the sum of its parts. Filmmakers do this by employing a tactic of storytelling evolved by early twentieth-century filmmakers called *montage.* They juxtapose and blend filmic elements—pictures, sounds, graphics, motion—to create idea associations that resonate with feeling in viewers. This happens because filmmakers depend upon viewers to create the meaning behind the blending of filmic elements. There are various levels of montage in any film, from simple to complex, but all involve viewers creating whole meaning by blending incomplete picture and sound stimuli.

When a filmmaker cuts from a shot of a woman looking off screen and then cuts to a picturesque shot of a beautiful ocean view, we blend the two into an association that we are sharing the woman's view of a landscape, even though we never see her observing the actual landscape. Instead, we take on her point of view and jump into her seeing perspective. The filmmaker has successfully lured us inside the story experience to occupy "the shells" of fictional characters.

Commercials use even more complex associations when they juxtapose short visual stories with their products. Such positive associations are designed to cultivate fresh and interesting consumer impressions that lead to product purchase. The filmmaker is making the product relevant to the viewer.

This relevance factor crosses into the classroom domain where teachers are involved in a similar process of engaging audiences. In this case, the teacher is making the lesson plan relevant to the learner. The difference is in the context of the "purchase" of the ideas. One method seeks consumption of the ideas for entertainment and profit, the other seeks retention for knowledge.

The Problem with Videomaking

While calling attention to the potential strengths of videomaking as a teaching resource, it is important to point out that it is by no means a "cure-all" or "magic bullet" for all possible learning situations. It is just a consistently misunderstood and often feared tool of learning.

The main problem with videomaking as a learning tool is its heavy technological component. With videomaking come cameras, wires, microphones, lights, tripods, videotape machines, and also the part of the process Ms. Bronson feared the most—editing.

These are all complications to an already complicated job: teaching K–12 children. I must admit that before I started making videos with K–12 classes, I had an overly romantic idea of what school was all about. In my distant memory of my own K–12 experiences, I realized I had smoothed the wrinkles and kinks and complications that go with the business of education—things like:

- Curricular mandates from districts and states
- Daily struggles for order and discipline
- Uneven abilities in learners
- Budget deficits
- Morale problems
- Complications of students with special needs
- Time deficits
- Parent problems and family issues

Every time I walked into a K–12 classroom, I walked out mentally and physically exhausted. It was extremely hard work just to get through a normal day. Try adding the complications and extra work that goes with the experience of videomaking and it's no wonder videomaking is *not* a part of every classroom in America, even if it poses tremendous learning advantages. Surviving the K–12 experience is a careful balance between meeting objectives and overcoming constraints posed by everyday reality.

This is certainly not to say that video cannot break through some or all of these complications. I have dozens of firsthand examples how it can! But it is important to understand the challenges that videomaking faces at the front door of real-life learning environments. The clear benefits of videomaking do not erase the challenges every classroom faces.

A Way In for Videomaking

The failures in K-12 videomaking I have witnessed all had one circumstance in common: They were without purpose—videomaking pursued only for videomaking's sake. This is not to say that every successful videomaking experience must be chained to deeply rational purposes and strict educational objectives. But if such experiences stray too far from the purpose and the context of the learning environment, they tend to become purely recreational, self-serving, and, in the end, not all that valuable, even to the children.

There is no contest in the debate between videomaking and learning. Learning is the horse, and videomaking is the cart. This is something that Ms. Bronson realized after her first video project:

> You don't want to make this a process that takes too much time trying to find information.... It should be readily available ... it was a topic I was very comfortable with, it was something I really knew pretty well, and it was a topic we spent a lot of time on as well. So it wasn't something that we spent just two days on. It was a major topic in our curriculum.

The Erie Canal unit was not just any topic to Ms. Bronson. It was a topic close to her heart that she wanted the children to truly learn and enjoy, *and* she was not quite satisfied with the way the lesson had been going in previous school years. It was a topic in need of a dose of learning impact. She hoped video could contribute to that impact.

News Class 104

Ms. Bronson and the students decided that the name for the program would be *News Class 104*, after the class's room number. Six reports would be produced in class by students working in small reporter groups of two or three. The reporter groups would make voice recordings of their reports that would later be put over pictures of what they were talking about. Students would also be responsible for finding appropriate pictures and videotaping them (right out of books) as full-frame television images. They used textbooks, library books, and class posters for most of their projects.

I would handle the editing of the video reports using university facilities, since at the time editing procedures were highly technical and not readily accessible to public school districts. This has certainly changed. At a later date, the edited reports would be rolled into a live news-style production produced by Ms. Bronson and her students at the university's television news studio.

It was also determined that the newscast would be hosted by four student news anchors. The anchors would write their own scripts and then read them on camera using the teleprompter—a device that projects the script text over a camera lens so the anchor does not have to read from a paper script.

Plans were also made to incorporate the class field trip, a boat ride on the Erie Canal and nearby water lock system. One of the reporting groups would make a story out of the field trip that would consist of images of the boat ride and interviews with field-trip participants. Other story groups could also shoot pictures of the actual Erie Canal if they needed them.

The culminating experience would be a half-day television studio session at the university where all program elements would be mixed together in the final newscast. The students had also decided to produce a "live" interview in the studio with a "Hoggie" (canal boat operator), creating the illusion that they were talking to someone from the late 1800s Canal Era. They would accomplish this by taping the Hoggie character in front of an 1800s background. They also made the footage of the Hoggie black and white, enhancing the notion that he came from an earlier time period.

It did not take long to decide who would be the director of the show. It would be the very person who was so worried about her inexperience in media production—Ms. Bronson. She didn't hesitate, even though directing television is no easy task. Ms. Bronson would call all the shots from the studio control room, and her students would operate all production equipment: cameras, lights, sound, graphics, video switcher, teleprompter, video recorder, and audio board. The program would be rehearsed several times, then videotaped for final compositing.

NEWS CLASS 104

(A VIDEO EXCERPT)

<u>NEWS MUSIC OVER SILHOUETTE OF FOUR NEWS ANCHORS IN FRONT OF TRADITIONAL NEWS SET</u>

BOTTOM THIRD CREDITS OF THE ANCHORS' NAMES APPEAR.

ANCHOR 1 (VAUGHN)
Welcome to News Class 104. . . . As you know, each week we talk about a different part of New York history and how it is important to us today.

ANCHOR 2 (CORY)
Today's topic is the Erie Canal. A part of our report today is a look at the canal era of the 1800s, and how the Erie Canal is still in use.

ANCHOR 3 (HANNAH)
We're going to have a series of reports including a live report to find out what life was like on the canal.

ANCHOR 4 (TAYLOR)
Our first report is the Erie Canal in its prime, and here's Jonell Jackson tracing the Erie Canal story.

GRAPHICS: "THE CANAL IN ITS PRIME"

REPORTER
Today our topic is the canal in its prime.

CUT TO MAP OF NEW YORK STATE WITH ERIE CANAL ROUTES.

REPORTER
This great canal connecting Buffalo on Lake Erie and Albany on the Hudson River once ran through the great city of Syracuse.

CUT TO PAINTED PORTRAIT OF DEWITT CLINTON.

REPORTER
The canal has a long and proud history starting with DeWitt Clinton, who was the founder of the canal.

CUT TO PHOTOGRAPH OF PEOPLE INVOLVED IN THE CONSTRUCTION OF THE CANAL.

REPORTER
Thousands of people helped to dig the canal over an eight-year time span. The digging was started on July 4, 1817, in Rome, New York.

CUT TO PAINTING OF BOAT UNLOADING OFF THE ERIE CANAL.

                    REPORTER
    On October 5, 1825, it was 363 miles long. That's
    when the canal era started. The canal became very
                     popular.

CUT TO PHOTO OF HUNDREDS OF PEOPLE LINED UP ALONG THE SIDE OF THE CANAL.

                    REPORTER
    During its prime between the 1820s to the 1860s it
    was the main way to travel in New York. Many towns
          and villages popped up due to the canal.

CUT TO PHOTO OF DOWNTOWN SYRACUSE IN 1800s.

                    REPORTER
      And some villages became cities like Syracuse.

CUT TO PHOTO OF TRAIN.

                    REPORTER
    By the 1860s, trains were used, and because they
       traveled faster, people started using them
          instead of the canal for passenger travel.

CUT TO PHOTO OF ONE BOAT TRAVELING DOWN THE ERIE CANAL.

                    REPORTER
    The canal started to fade away by the 1900s and
       the canal era ended. So that's the story of
         the canal era, now back to you Taylor.

CUT TO TAYLOR IN STUDIO.

                    TAYLOR
          That was excellent Jonell, thanks.

[PROGRAM CONTINUES WITH FOUR MORE REPORTS AND A LIVE INTERVIEW WITH A HOGGIE]

Ms. Bronson was actually quite a natural when it came to directing, and her crew was fabulous as well. Part of it had to do with the fact that they *had* to do it, because they committed themselves to the project. None of them had ever done anything like this before, but it didn't matter. They didn't have their own production staff to do it for them, so they had to step up and perform in their roles and get the job done—and they did.

The most important part behind the experience was that they had a learning objective that transcended their video work—they had something to say. They had a clear objective to attain, a show to produce, and it was all theirs. The topic was history and it required a great deal of research and preparation in order to pull it off. They were learning their social studies curriculum and research and study skills in the context of their media production. These experience-centered lessons were sticking. Ms. Bronson later shared her perspectives on the use of videomaking in her curriculum:

> I think that it motivates kids a lot. I think that it gets kids to want to do more on this topic. I noticed kids when we said "OK, we're going to study the Erie Canal," they were like, "OK, well whatever . . ."
>
> And then it was, "OK we're going to do this [video] project," and then it was like "Oh! Well, now we have to do this because we want to be part of this project!" So they were a lot more motivated. . . . They had to do their part. They had to then be part of the team and that was another big plus. Even though we were working in small groups, we were small teams and then we were within a bigger team. And so it was really a bonding experience, really, it was really an amazing experience.
>
> So I think that then you get so many of these little extra benefits from it that it's worth trying. It doesn't have to be extremely sophisticated. I think that just putting on the camera . . . the kids are just fascinated by it.

After our videomaking experience, the benefits were clear to Ms. Bronson. Ironically, they had nothing to do with her fears of media or even media in general: the benefits were all about her main objective—enlightening fourth-grade students.

Making Sense of Video and Curricula

The question that has continually emerged throughout my K–12 videomaking activities is *not* whether video works or not. It clearly does. Videomaking stimulates the learning environment in the classrooms it is used in. The question has become, "Why does it work so well?" Diana Green elaborated on this in the context of her second-grade class:

> If I wanted to teach a social studies concept about history, I would be apt to read a story about a particular person, place, or time in history. The story's organization and the familiar concepts to the kids present them within the familiar concepts and organizational themes. They can take the content and understand the objective. If I want to talk about what it was like for the Pilgrims, I read a story about a little kid their age in Pilgrim times, and the kinds of things that they did. To translate that to video media, the words are the same. The pictures are the same. Sometimes it's animated. Sometimes it's not. The level of attention is about a hundred times more involved than if I were reading it to them. Whether it's good or bad or indifferent or whatever, I think we need to take advantage of it. It's a motivating factor.

Just like *watching a video* seems to step up a student's involvement in a subject matter, compared to a printed account, *making a video* seems to step up involvement, only more. This is because it's not just content anymore, it's content they have produced.

Why Videomaking Helps Children Learn

In his studies with British schoolchildren making videos about their local environments, David Gauntlett, like me, seemed to find the videomaking-curriculum connection in every class he worked with.

> Children became eloquent and energized about subject matters they knew little or cared little about beforehand [on students at Royal Park].

> Making video enlivened concern of environment in students who didn't seem to care about it before [on students at Blenheim School].

> As at Royal Park, although most of the children had been far from committed environmentalists in a general sense, when faced with their local surroundings they became enthused with ideas and complaints, and apparently came to see both the good and bad aspects which they had not previously noted or discussed [on students at Burley St. Matthias].

> The video project appeared to bring the subject of the environment "alive" for the children concerned [on students at Brudnell]. (1996)

Perhaps one reason why videomaking helps children learn might be its affinity for experience-based learning, as Diana Green realized after her class made the *Pollution Rangers* video.

> I think that the kids not only learned their content, because all year long any discussion of pollution or littering or anything, their eyes perked up and they knew what they were talking about or they would correct people. Ms. Michaels [her assistant teacher] had this terrible habit of throwing her soda can in the trash after lunch every day and I don't think one day went by that somebody didn't get it out of the trash and give her a lecture about it [properly placing it in the recycling bin]. They just really knew their content.

Videomaking made the lesson plan larger than just the lesson content. It was somehow increasing students' level of involvement in the subjects it was connected to. To James Marshall, technology like video opens up new learning opportunities and enhances retention.

> Engaging the learner through text and visuals has proved an effective means to enhance retention. But placing the learner in the middle of the content and responsible for making decisions and acquiring knowledge takes learning one step further . . . learning by doing results in new knowledge and retained knowledge. . . . The ability of media to engage the learner, activate emotional states, initiate interest in a topic, and allow for absorption and processing of information shares a direct relationship to the potential that learning will occur. (2002)

Such engagement also improved student behavior in Ms. Green's view.

> Whenever kids are doing something they are invested in, behavior problems are really pretty minor—when they're involved in choosing, making some decisions about their activities. So whenever we see that, they were putting together a lot of strands of things they had learned from social studies, science, health, language arts—lots of different kinds of curricula were coming together to make this movie. And because they had a handle on this curriculum and they were using it for a specific purpose, it's very brain-friendly learning. I've learned that kind of activity really negates behavior problems.

There are plenty of theoretical explanations for the phenomenon of videomaking's curricular success. In the book *Literacy in the Television Age,* S. B. Neuman relates two helpful theories: dual coding theory and the theory of instructional means. Dual coding theory, explained Neuman,

> proposes that two [representational] systems are actually at work, a verbal system specializing in processing linguistic information in various modalities, and a separate nonverbal system for spatial information and mental imagery . . . a child looking at a picture might also engage in covert verbalization. When a concept is registered in both memory systems, it is said to be dual-coded. Paivio argues that dual coding is more likely to occur with the image than with the verbal system. Since two memory traces are better than one, the dual-coding theory suggests that memory for visual imagery may be superior than memory for words. (1991)

The theory of instructional means suggests that media open up new windows on learning that traditional means simply have never had. This theory, Neuman writes,

> suggests that a medium's symbol system or code not only conveys knowledge but rather cultivates new skills for exploration and internal representation (Clark and Salomon, 1986). In the school setting, the skills acquired are based on the medium of text. As a result, Olson reports "when children are taught to read, they are learning both to read and to treat language as text" (1977, p. 103). Television, employing its unique symbol system, will yield a different set of skills in acquiring new knowledge. (1991)

Some have suggested that video could, perhaps, stimulate a different portion of the brain than other more traditional learning activities. Marshall (2002) characterized some exciting developments in brain theory, including Howard Gardner's concept of multiple intelligences:

> The relative strengths and weaknesses among and between these intelligences dictate the ways in which individuals take in information, perceive the world, and learn. Most traditional textbook approaches to teaching a particular subject favor a linguistic or narrative approach. Such an approach will fail to reach those who may respond better to an artistic or naturalistic depiction of the topic. In addition, it also fails to develop those other neural connections and pathways and further enhance those intelligences. This is where technology-based approaches incorporating video and audio (in other words, multimedia) allow education and, in effect, learning to reach more students and provide more opportunities for neural development and learning.

When I think back to my first days of using videomaking to teach media literacy concepts, it's almost laughable now for me to think I had no idea the technology could have such a profound impact on traditional lesson plans as well.

Extracurricular Benefits of Videomaking

By the time I had reached Ms. Bronson's fourth-grade class, I was fully aware of the relevance factor and other curricular connections. But as usual, just when I thought I had it figured out, I realized something else exciting about the videos we were making that I had never considered before: their shelf life! Ms. Bronson explained this:

> We show it every year. In fact, the other teachers from the team also want to borrow it and they show it before they go on the Erie Canal. It's part of their Erie Canal lessons . . . there were certain things that were caught on camera that really helped them understand a little bit more.
>
> They saw pictures of the children, they were on a boat trip on the Erie Canal. And so they saw a little bit more about what that was about, and they saw a boy dressed a little bit more like a Hoggie and so they thought, "Oh that's very interesting!" And of course they were captivated by the technology, of "How did that happen?" But it reinforced a lot of the terms: we talked about Hoggies, we talked about Clinton's ditch, they heard about the locks and so all that was sort of, "Oh, that's what we're studying!" and it sort of came together for them.
>
> And so . . . it supplements very well . . . and it just goes hand in hand. . . . It's hard to get videos that go with a certain topic sometimes, or they're too hard, they're too high level and this was at their level, they really understood it, because kids wrote it.
>
> It's perfect!

Note

1 Gestaldt—ge·stalt or Ge·stalt, n: a set of elements such as a person's thoughts and experiences considered as a whole and regarded as amounting to more than the sum of its parts (*Encarta World English Dictionary* 1999).

CHAPTER 8

VIDEO PLAYING FIELD

The Miracle of "Must Pollute!"

We needed a way to show the effect the two-headed alien toxic pollution monster had on Earth children when they cast their evil pollution spell. The script called for the children to turn into wandering zombies in search of something to pollute—air, land, water, whatever.

Hannah had the solution. "Arms straight out, eyes rolled up under her eyelids, and a methodic march forward to the chant, 'Must pollute! Must pollute!'" Just standard zombie behavior, really.

POLLUTION RANGERS OF THE SEVEN SEAS

(A VIDEO EXCERPT)

<u>THE TWO-HEADED ALIEN TOXIC POLLUTION MONSTER IS WATCHING MS. GREEN'S CLASS USING ALIEN EAVESDROPPING TECHNOLOGY THAT BEAMS IMAGES AND SOUNDS BACK TO THEIR SPACESHIP HOVERING SOMEWHERE ABOVE EARTH.</u>

THE TWO HEADS ARE HAVING A CONVERSATION ABOUT THEIR EVIL PLAN TO GET RID OF ALL THE TRASH FROM THEIR PLANET, JUNKO.

BECAUSE ALL OF THE INHABITANTS OF PLANET JUNKO HAD NOT CARED ABOUT THEIR ENVIRONMENT, GARBAGE AND POLLUTION WERE EVERYWHERE—IN FACT, THERE WAS NOWHERE LEFT TO PUT IT!

THIS IS WHY THE TWO HEADS WERE SENT TO THE PLANET EARTH. EARTH WAS TO BECOME THE NEW GARBAGE DUMP FOR THE PLANET JUNKO.

          MONSTER HEAD ONE
Ha ha ha ha! Now it's time to unleash our evil plan.

          MONSTER HEAD TWO
      What are we gonna do?

          MONSTER HEAD ONE
All we have to do is to make them think that they *can't* make a difference, that it's no big deal if they pollute. Someone will pick up after them. One little piece of litter isn't going to hurt anything or anybody. When they think that, we will have all the power we need!

         MONSTER HEADS IN UNISON
        Ha ha ha ha ha!!!!

CUT TO LUNCH ROOM, THE NEXT DAY, CUCKOOVILLE ELEMETARY SCHOOL. HANNAH IS EMPTYING THE CONTENTS OF HER TRAY INTO A GARBAGE CAN. ONE OF THOSE CONTENTS IS A RECYCLABLE CAN, WHICH BRIAN NOTICES AND CALLS ATTENTION TO.

              BRIAN
Hannah, you're supposed to put your recyclables in the recyclable container.

             HANNAH
That's OK, it doesn't really matter.

AS THE WORDS COME OUT OF HANNAH'S MOUTH, THERE IS SOME SORT OF NOISE, LIKE AN ELECTRIC JOLT. HER BODY SNAPS

SUDDENLY LIKE SHE HAS BEEN SHOCKED. HER EYES ROLL UP
INTO HER HEAD. HER ARMS SPRING STRAIGHT OUT IN FRONT
OF HER AND HER VOICE BECOMES MONOTONE AS SHE CHANTS
WITH ABSOLUTELY NO EMOTION OR HUMANITY:

                  HANNAH
        Must pollute! Must pollute!

HANNAH WALKS IN A ZOMBIE-STUPOR, STRAIGHT PAST BRIAN
AS IF HE DOESN'T EXIST. BRIAN'S MOUTH IS WIDE OPEN IN
SHOCK.

This scenario was repeated with several other children (including Mrs. Irwin's entire second-grade class!) and even some adults. The outcome was the same, instant zombies marching to the chant "Must pollute!"

The "must pollute" chant was one of the most memorable aspects of the videomaking experience, especially to the children who watched it. The last scene of the movie showed Corrigan's vice principal marching in zombie-state past one of the Pollution Rangers who realized the Rangers could never rest when it came to taking care of Earth.

After watching the screening of *Pollution Rangers* at the school, my youngest son, Elan, who was a year-and-a-half old at the time, seemed to especially identify with the act, as did Samantha. Samantha was part of Corrigan Elementary's Inclusion Program that mainstreamed students with physical and learning disabilities with the general school population. She was diagnosed with autism and required constant one-on-one supervision in Ms. Green's second-grade class. Ms. Green related the miracle of "must pollute" with Samantha.

> One day Elan came in [to his older brother's class] for show-and-tell and did his "must pollute" routine, and that was the biggest hit, and you can't even imagine. I'm sure everybody will remember that day when Elan came in and did "must pollute!"
>
> And then the day Samantha did "must pollute," because Samantha refused to take part in the movie, and one day she came up to me and she said, "Ms. Green, Ms. Green..." [She makes "must pollute" motions with her arms stretched out mimicking Samantha's actions] "Must pollute! Must pollute!"
>
> For her to have gathered that language and to have used it, as a handicapped student without language, was really amazing. I think the effects were always there. The kids talked about it, it was a solid given, something that they could put in their schema and relate to in many different ways.

The "schema" that Ms. Green was talking about was the particular way that writing and ecology were approached as subjects of the class. Using the experience of videomaking allowed them to experience the lesson plans in very different ways and these "ways" had unexpected results, particularly on those who had learning problems.

First-grade teacher Sally Beechwood described this phenomenon as "a leveling of the playing field."

> I think there's a little bit of everything. First of all it's fun for the kids. And you want to make school fun, especially at that age.... Just reading books is fun. A lot of things are fun. But, you know, it kind of levels the playing field for them. Because everyone can be part of the movie and you don't have to be the best reader and you don't have to be the best this or that, so it levels the playing field.

What teachers and I found time after time was that videomaking seemed to noticeably stimulate students who were not usually thought of as the most able or intelligent in the class. Videomaking consistently "awakened" students who had been labeled lower performers in classes. Even students with severe learning disorders seemed to come to life in some way.

The teachers and I were finding little correlation between a child's "classroom intelligence" and their "videomaking intelligence." Often, as Ms. Green relates, the most visually inclined students were *weaker* performers in traditional subjects and classroom activities.

> Many of the children seemed to have a personal sense of achievement from it that sort of made them come out of their shells.... Quieter children seemed to be much more a part of the activities in the classroom.

> So many children in our class perceived themselves as not being able to do things, kids who may not be academically where their peers are. Some of them had major parts in the movie, and that was something that they felt good about.... People all over the school saw them doing this and they were very, very proud of that.

This was not a finding reserved for Ms. Green's class. It seemed to surface in every classroom at every level in different ways. It also paralleled some of the latest research on how brains learn.

Gardner's theory of seven (now eight) intelligences connects to the kinds of experiences that I was having with young learners. Gardner moved well beyond some long-standing explanations of intelligence—for instance, IQ measurements and right/left brain functioning—to suggest a much more complex process behind functioning. He saw learning and intelligence as a multidimensional process described seven distinct intelligences that operated within this process:

1. Linguistic intelligence—sensitivity to language and the relations among words
2. Logical-mathematical intelligence—abstract thought, precision, counting, organization, logical structure
3. Personal intelligence: awareness of others
4. Personal intelligence: awareness of oneself
5. Spatial intelligence—observation, mental images, metaphor, gestalt
6. Musical intelligence—sensitivity to pitch, rhythm, timbre, the emotional power and complex organization of music
7. Bodily kinesthetic intelligence—control of one's body and of objects, timing, trained responses that function like reflexes

These intelligences did not operate in separate vacuums, according to Gardner. To the contrary, they flexibly connected and complemented each other with no prescribed pathways among them. Although independent, they tend to function together, adding dimension to learning.

Over the course of my K–12 videomaking experiences, there were plenty of examples of how videomaking activities seemed to activate dormant or little-used thought processes in children.

The Case of Shanara: Equal Footing

Shanara was cast as one of the Pollution Rangers of the Seven Seas. Her character name was "Atlantic," and she not only had some problems in pronouncing the word "Atlantic" but also in simply remembering it. She was surrounded by children who had no problem at all remembering or pronouncing their names. Ms. Green elaborated on the case of Shanara:

> She didn't know all her letters when she came into second grade. And . . . there are several letters that she doesn't pronounce properly. She has usages that aren't correct. Everything is hard for her. The demands of the second-grade curricula are hard for her. The demands of the first-grade curricula would be hard for her. And she's somewhat developmentally delayed.
>
> If you had any idea how many times we had to tell her what her movie name was . . . that scene in the office. . . . We probably had to do it three times, just because when we got to her she'd say "What's my name again?" But when the movie came out, she was on there doing it right. The mistakes were irrelevant. There she was being successful at something. Now she can watch over and over and know that she can be successful. And she's

forgotten how many times she asked me what her name was. If you asked her today, she'd know what that name was. And to Shanara, she was doing the same thing that Meredith was and what Mary was. And she knows the answer. The movie put them all on an equal footing.

The Case of Carly: Counting Backward

The "labeled disadvantaged" seemed to genuinely thrive and in ways outperform the "non-disadvantaged" in unexpectedly positive and eloquent ways.

—David Gauntlett (1996), on students from Little London

It was already too late by the time I found out Caitlin had a severe learning disorder. The damage was already done—not bad damage. It was quite good, actually.

It happened in the control room of the university television studio when we were taping the segments of *The Culture Project*. We were recording students talking from the various countries they had been transported to by a defective time and space machine. We situated them in front of a green wall and substituted the green color with a background picture representing the country they were in. The segments would later be edited into *The Culture Project*. This production required a great deal of coordination among the students and I was relying on Caitlin to keep things in order. She was put in the role of assistant director in charge of starting the videotape machines, counting down to air time, and keeping track of segment lengths. Caitlin not only did a fantastic job in this role, but she was also the main writer of the movie, always coming up with great script ideas and thinking of solutions to creative challenges we had.

After we had finished recording all of the segments, one of the assistant teachers pulled me aside, somewhat shaken by something she had seen. She informed me that Caitlin had a severe learning disorder and much of it had to do with counting. She had a lot of trouble counting numbers, yet here she was in charge of timekeeping. She had no trouble whatsoever in counting numbers as an assistant director, and on top of that she was counting backwards! 5-4-3-2-1 ... something her special education teacher had never seen her do before.

The Case of Meredith: Becoming a Diva

The video project had brought out abilities in the children, which in some cases in particular, had previously been less than conspicuous.

—David Gauntlett (1996), recounting a teacher's observations of students at Blenheim School

Meredith Townes was a very bright student in Ms. Green's class, but terribly shy. One of the reasons Ms. Green had cast Meredith as one of the Pollution Rangers—in fact, the Pollution Ranger with the most lines—was because she hoped she could get Meredith to talk more. She recounted Meredith's progression from *girl reluctant to raise her hand* to *class diva*:

> She would respond when called on, but she wasn't seen often . . . and she was such a star in the movie, and that seemed to me to be very indelible.
>
> The parents are so proud of the kids creating it, as well as acting in it. Mrs. Townes [Meredith's mother] was surprised that Meredith was as outgoing in the movie as she was, because Meredith's pretty quiet. And she related a story to me about a month or so later: Meredith had been playing with some paper dolls with some friends and she organized them all into a show. She was the director and told everyone what to do, what to say, where to stand. And she said she never had seen Meredith act like that before. . . . She's always been very quiet and shy and she really did change a lot this year. She's been very vocal. It stayed with her.
>
> Also, Mrs. Townes teaches dancing and Meredith is in a couple of her classes and she told me early on in the year when I told her, "I know Meredith knows things but she just doesn't talk."
>
> She said, "She doesn't talk in dancing either." But by the end of the year that had really changed. As a matter of fact, by the end of the year, I must have said fifty times during the month of June, "Meredith, stop chattering, it's my turn to talk!" Which in September, I don't think I would ever dreamed of saying that to Meredith.

The Case of Poison and Syrac: Companionship through Evil

Wendall and Jenny were cast as the evil duo: Syrac and Poison. Their mission was to rid the world of all books so that Syrac (who couldn't read) could be the all-knowing power on Earth. They were the antagonists in this high school series *Imag-A-Book and the Read-A-Lot Gang* and therefore always up to no good.

In real life, Wendall and Jenny were near opposites. Wendall was, broadly speaking, a sophomore nerd. Jenny was a popular senior. The two never would have spent time together had they not been thrust into their performance roles for the TV series.

They became very close over the course of the production and remained close friends after the show was finished. There was something about the alternative world videomaking provided that allowed space for people to transcend their everyday roles and associations. Many of the teachers I worked with seized this opportunity to affect class chemistry. Diana Green was notorious for this when she cast her movies. She liked to present alternative relationships—to put the quiet students in loud roles, the powerful in powerless roles.

Teachers didn't have to orchestrate this effect in their video projects. It happened naturally as a rule. A special education teaching assistant explained it best when she commented on the making of the third-grade movie, *The Culture Project*.

> When I saw them paired up on the screen for each country, kids that were together were not kids that often played together in here, and I just thought it was a real nice way of getting kids to socialize better. I mean, it wasn't even part of the unit in terms of the social studies.... Everybody was very accepting of each other.

In the same way that videomaking encouraged different levels of thought in learning, it seemed to encourage uncommon social interaction as well.

The Case of Neddie: Neddie Being Ned-D

> Video work therefore offers non-white children an opportunity to create a text involving themselves and their culture which is more likely to side-step teachers' concerns about linguistic "correctness," and which can record forms of expression directly, without there being a need to have them written as words on a page. Children who have limited abilities in written English become able to demonstrate their creativity and intelligence, as can those whose forms of communication differ from the school's norm of "standard" written English.
>
> —David Gauntlett (1996), on students from Beckett Park

Although much about Neddie was unique, what he represented as a "problem student" was far from unique. There were usually one or two "Neddies" in each Corrigan classroom. They were equally distributed, according to Diana Green.

> At the end of June, when classes are reorganized for the following year, some kids are obviously more needy than others, take more teaching time, take more time for behavior needs than others. And those kids are kind of earmarked, because you don't want them all in one classroom, both for themselves and for their teacher's sanity. So Neddie was one of the kids that was obviously earmarked and he landed in my classroom. And so, of course, we have all summer to think about, "Oh dear, how are we going to deal with this one or that one?"

But, everyone in the school seems to know Ned. So, this is a kid you know when he walks in the door day one. He's very visible. When you go to the auditorium for a show, Neddie will be pulled out halfway and he does it in such a noisy way. And of course, second-grade teachers look at first-grade students and third-grade teachers look at second-grade students and I'm sure what's going through their mind is, "Oh I hope I don't get that one!"

But you know, ... everyone in the school knows who Neddie is.... That's just the kind of kid he is. So, the first thing that struck me about Ned was that we wrote our names on a piece of paper and he wrote his name "NEDD" and I said, tell me why you spell your name that way and he said, "My name is Neddie.... Here it is. Last year I just wrote NED, but this year I'm going to write Neddie—NED-D." He sounded it out. And I thought, "Well, he can't be all stupid. That's a pretty good thought process!" We have to take small gains where we get them in second grade. And he has a thought process like this. He marches to a different drummer and he thinks! It's just that he never comes out with the same bottom line that you would or I would. And half the time, you think he's really off the wall—you have to step back and remember that Neddie's process is a little different.... He's below average in overall intelligence. But he does have these places were he can shine, where he can do things well.

Neddie shined with the videomaking project in Ms. Green's class because it provided an arena for his often unrecognized and unappreciated strengths.

Learning More with Videomaking

In each of these cases, videomaking opened a doorway to learning that didn't seem to be there before. What can we do to exploit the expanded learning possibilities that the experience of videomaking offers? One possibility is to apply it to problems or difficult subjects.

There is a tendency for video to "get at" topics that traditional approaches are not working effectively with. The third-grade bomb-threat problem was a good example. This was an event that a standard approach—for instance, direct teacher-to-student explanations and discussions—did not seem to work effectively with, especially given the frustrating persistence of the problem. Getting kids to explore the issue using the more abstract expression of videomaking seemed to fit with the complicated nature of the problem.

Ms. Bronson's use of videomaking to get students to identify more with the research process was another example. Her standard approach to the subject was not delivering the impact she had hoped for. When we folded the research process into the highly action-based and aural-visual experience of putting a newscast together, it enlivened the process for the students.

Alternative Learning Opportunities

> Children whose first language was not English were able to display their abilities to the full in this non-written form.
>
> —David Gauntlett (1996, 115), on children at Brudnell

Though English was my son Logan's first language, you would certainly think it wasn't when it was time for him to read his history textbook. I couldn't really complain, though. He was getting straight A's all the way around, and his teachers raved about having him in their class. What more could I ask from him?

But Logan wasn't happy about the situation. To him, he was not learning. He could get great grades, but was having trouble learning. It centered on his advanced placement (college-level) Global History course. The problem was that he couldn't read. Whatever way he approached the reading, it was all turning into "fuzz" and making him fall asleep, one or two paragraphs into the reading. Part of it was his inability to find purpose in the reading. "Why do I have to read this stuff? What's the point?" were two of his most-used phrases at study time.

The other part was that even if he put these questions aside, he could not follow the words to any sort of point, outcome, or learning objective. He would simply doze off. I wondered if he might have a learning disorder.

I tried to explain that reading for studying was different than reading for pleasure, but that didn't make him feel any better. So I brought him in to see a learning specialist, Marlene Blumer from Syracuse University's School of Education.

It didn't take her long to diagnose the problem. She gave him two tests, once to diagnose his inherent learning style and another to interpret his intelligence on a multiple intelligence scale.

What she found was that Logan was a typical high school student of the twenty-first century. He didn't have a learning disability. He was just not a reading/writing (linguistic and logic centered) learner. He was instead a kinesthetic, visual, and aural learner.

He was conditioned for learning by doing, learning by seeing, learning by hearing... *not* learning by reading and writing. Marlene came up some with some approaches to learning Logan might try, given his orientation to learning—things that focused his studies on his learning strengths. In the end I was happy, for two reasons: one that there was nothing really "wrong" with Logan, we just had to rethink his learning strategy. The second reason was my realization that young learners of the twenty-first century need fresh approaches to learning... and videomaking can help with this.

Given the abundantly visual world children are raised in—with television, video games, Internet, DVDs, and now even i-pods and cell phones—challenges to traditional learning methods are likely to get more intense. Traditional methods must evolve to the needs of learners at some point. Videomaking could be a potentially helpful tool in this evolution.

Altering the Learning Environment

Yet another thing to consider is the interesting effect videomaking can have on power relations, roles, and labels. Ms. Green demonstrated that she could effectively alter the roles and labels of her students by portraying them in ways that were the opposite of their everyday appearances. This tactic could constructively alter students' perceptions as they relate to classroom dynamics and self-esteem.

In her second movie, Ms. Green decided to apply videomaking as a means of behavior modification. She was having some problems with her students not realizing the line between fun and seriousness, so the class wrote and produced a movie with this theme. It was called *Comes a Time* and it was about how students had to realize that learning required hard work and focus on their part, and this was a good thing. It didn't mean they couldn't have fun. It just meant they had to understand the point at which the two objectives worked against each other. The motto repeated throughout the movie became the rally call for the rest of the school year:

> The first time it's funny.
> The second time it's boring.
> The third time you're *out of here*!

The movie served as a reference point for any questionable activities on the students' parts for the rest of the school year.

Making a Difference

Videomaking represents a true opportunity for students to learn more. Sometimes this means new content, but it can also mean new angles on existing content. In line with Gardner's (1983) theory of multiple intelligence and even dual coding and instructional means theories (Neuman 1991), it seems appropriate that there be dimension and plurality in learning experiences. And based on the latest brain research, it appears the brain is actually wired for it.

Diana Green certainly has picked up on this development in learning. I recently dug up an e-mail she sent me after our first movie. I had designed a video-scripting assignment as an experiment to see how much her students remembered about putting visual ideas together. Ironically, the experiment failed because they really didn't remember much at all. Perhaps it was because we were alternating so much between script and screen, or perhaps they were just exhausted from making the movie. But in the end I think Ms. Green zeroed in on the truest value of our videomaking experiences.

> I sent the movie-script assignment home with [your son] this morning. Not everyone turned them in, but you have a healthy sample. Neddie had no idea where his went, as usual, which I thought was too bad.
>
> I hope you realize that you have me hooked into this video literacy idea, and you may not be able to escape making another second-grade movie. Can Vaughn stay in second grade for a few years?
>
> Seriously, thanks for everything, a million times: the movie, the help in the classroom, your good humor, your all-nighters, etc.
>
> It makes a big difference for kids, and some of them need every advantage.

CHAPTER 9

THINKING IN VIDEO

> We are what we are able to see, even more powerfully and profoundly than we are what furniture of ideas we have stocked in our heads.
>
> —Susan Sontag (Barnhurst 1987)

IT SEEMED Mary Catherine Catrell was born to be a fifth-grade teacher. When I walked into her class, it was clear that she had a keen understanding of the fifth-grade mind. The students in Room 204 absolutely adored MCC—their agreed-upon pet name for Mrs. Catrell. MCC believed the fifth-grade experience had to transcend the curriculum to nurture the changes they were going through. To her, it was as much about environment as it was lesson plans.

> You set up a learning environment for the kids, and I really like it to be kind of playful and fun. I don't think you remember kindergarten your whole life, but you remember fifth grade. It's like a turning point. They come in, they're really babies and then by the time the end of the year comes, you know they've really grown. They have opinions about things, they're maturing a little bit. I mean, it seems funny because they're only ten and eleven and twelve years old, but they feel like they're really adults by the end of the year.

MCC worked very hard to make her fifth-grade experience unforgettable. This is one of the reasons she agreed to make a video with her class even though she was terrified by the idea of it:

The first video ... really, I was so intimidated ... and I just thought, "I don't know how to do this!" And what happens is the kids know how to do this! And they love it! It's like a different venue for them.

Visual Inequity

It is beyond cliché to say that we're living in a visual world, with its hundreds of television channels, movie-plexes, the Internet, video games, DVDs, just to name a few of the more prominent sources of imagery in our lives.

Despite the visually intense landscape of everyday life, the school systems I worked in continue to emphasize traditional text-based teaching methods in their classrooms.

It doesn't help that there is a tremendous gap between teachers and students in visual knowledge. The kids I worked with at Corrigan Elementary were far more visually intuitive than their adult teachers. The teachers didn't tend to know how to harness this intuition in their learning environments, and were therefore missing out on a valuable teaching resource.

Part of the gap was due to children's early, natural, and voluminous exposure to visual media. This experience that children arrive at school with is a considerably untapped learning resource, as Paul Messaris explains:

> Unlike conventions of written language or, for that matter, speech, pictorial conventions for the representation of objects and events are based on information-processing skills that a viewer can be assumed to possess even in the absence of any previous experience with pictures. (1994)

Another reason might be that visual thinking is very much like language and kids are more open minded to alternative discourses at younger ages. It is considerably more difficult for teachers (old dogs) to learn other languages (new tricks) with their comparably ingrained knowledge bases.

Videomaking involves translation from three-dimensional ideas to two-dimensional ideas. Filmmakers must reframe every idea or experience or risk miscommunicating their ideas. Converting to two-dimensional expression is not unlike learning to walk or talk from scratch.

In effect, what we have is a "blind" TV generation teaching the intensely visual children of the TV generation. The TV generation is, compared to its children, visually illiterate. How then can the TV generation utilize visual teaching resources if they are visually illiterate? This creates an interesting contradiction: the "blind" leading the "seeing." It will take a new breed of teacher to implement visual thinking into curricula, not unlike those who integrated computer technology into a once computer-illiterate school culture.

There is certainly some mystique involved in such a leap of faith. The fact is that participating in visual curricula thrusts the children beyond teachers and parents in much the same fashion that computers and computer games have. This is clearly evident if you watch a typical adult try to compete with a typical child in a typical video game. Children are much more agile and experienced mostly because of the time they spend "practicing," and adults are not likely to catch up to them as far as game skill goes.

This point was also evident in a conversation I had with a retired Air Force general. He was talking about his lingering urge to fly. He mentioned that flying was a whole new thing now with technology advances. Pilots must now do so much more than when he was a fighter. There was much more for them to learn and subsequently worry about. He couldn't imagine doing all that extra technology, outside of flying, himself. There is a similar generational dynamic at play in visual thinking and expression.

But there is a saving grace in all this, even in flying. As we saw in the last chapter, success with visually enhanced curricula is not as much about a teacher possessing visual savvy or technical know-how as it is the will to illuminate and the courage to find effective teaching methods. Even in flying, if you strip away the technology, there is an overriding mission that any technology must be connected to, or it will fail.

Visual Thinking in an Information Age

The challenge of creating more visual curricula is similar to the challenge David Berlo identified in adapting education to the information age. As a former university president and communications scholar, he declared that the human brain had officially succumbed to the early information age. Berlo related,

> For the first time in human history, two related propositions are true. One, it no longer is possible to store within the human brain all of the information that a human needs; we can no longer rely on ourselves as a memory bank. Second, it no longer is necessary to store within the human brain all of the information that humans need; we are obsolete as a memory bank. . . . Education needs to be geared toward the handling of data rather than the accumulation of data. (1975)

He was, in effect calling for an educational revolution, an overhaul of the very underpinnings of traditional learning. There have been both inroads and obstacles in addressing Berlo's challenge over the past thirty years.

Inroads

Berlo's ideas were not far afloat from Marshall McLuhan's. In *Understanding Media: The Extensions of Man,* McLuhan explained: "We see ourselves being translated more and more into the form of information, moving toward the technological extension of consciousness" (1964).

In his 1995 book *A Celebration of Neurons: An Educator's Guide to the Human Brain*, Robert Sylwester described this extension of consciousness in more genetic terms:

> Because biological evolution proceeds much slower than cultural evolution, we're born with a generic human brain that's genetically more tuned to the pastoral ecological environment that humans lived in thousands of years ago than to our current fast-paced urban electronic environment. Our curiosity and inherently strong problem-solving capabilities allowed us to develop such tools as autos/books, computers/drugs that compensate for our body/brain limitations—and very powerful portable electronic computerized instruments are now rapidly transforming our culture. (1995)

Further developments in technology, Sylwester explained, actually extended the range and capacity of our brains:

> Consider the organizational functions of our brain and the great variety of technologies we have developed to get beyond our limitations—technologies that regulate things, enhance sensory input and motor output, and remember/analyze/organize information for decision-making. We can almost think of technology as another layer of the brain on the outside of our skull that extends performance beyond the limited capabilities of the brain on the inside. (2003)

Gavriel Salomon (1993) used the term "distributed cognition" to explain such an intellectual migration outside of the human brain. This involved everything from notes stored on notebooks to interdependent knowledge in groups or corporations. Training in methods of handling information outside the brain, Salomon explained, would prepare students to work in all kinds of unpredictable environments and encourage them to learn about the world on their own. The emphasis here was on tools to handle information, as opposed to emphasis on domain knowledge and stored facts.

Obstacles

In "A Plea for Media Literacy in Our Nation's Schools," David Shaw summarized the problems surrounding reform efforts in education. He saw reforms as falling outside the "formal cannon" of status quo standards: "Education in America is very structured, resistant to change.... With school budgets tight everywhere, it's difficult to introduce programs or classes or hire more teachers to administer those programs and teach those classes" (2003). This is why he and others suggest the solution is to incorporate new techniques in existing classes—a "cross-disciplinary" approach.

Picture Worth a Thousand Words

Enhancing visual thinking represents a way of dealing with the challenge of handling surplus information flow that David Berlo presented. Given the inexhaustible flow of information in present times, it is increasingly necessary to devise ways of handling information rather than ways of storing it in our limited memories. Visually based activities like videomaking offer rich contexts that frame content in ways that are close to actually experiencing it.

This is something that MCC noticed in her two moviemaking experiences. She coined it "the movement piece."

> It's not just reading it, writing it, drawing it . . . it's the movement piece. In the poetry video, one of the things they start out with, the kids are like [moves side to side mimicking the opening dance number in the second poetry movie] doing this dancing thing. . . . They need to move, and there's something with the video . . . they're all moving and playing and they love to direct each other and I think it helps them to remember. They've got their case, you know how they're going to make this come together so they have to think of all the pieces, and they're doing that research and they're putting it together, and I think it's going to help them remember it. . . . It's just a different way to express themselves.

The Poetry Videos

One of the most dominant themes of MCC's fifth-grade experience was her poetry unit. Yes, it was an official part of the fifth-grade curriculum, but something MCC felt a particular closeness to. Part of it was the pure challenge of taking poetry on.

THE POETRY VIRUS
(A VIDEO EXCERPT)

<u>WE HEAR DRUMS BEATING WITH NATURE SOUNDS AND FADE INTO A SHOT OF TREES BLOWING GENTLY IN THE WIND.</u>

<u>DISSOLVE TO SHOT OF THREE FIFTH-GRADE STUDENTS DANCING SIDE-BY-SIDE ON A PLAYGROUND (A GIRL BETWEEN TWO BOYS) SWAYING IN ELEGANT SYNCHRONICITY TO THE HEAVY PERCUSSION BEAT.</u>

THIS IS A DANCE THEY CALL "THE SNAKE." WE HEAR THE
CHORUS OF CHILDREN'S VOICES READING A POEM TOGETHER.

> CHILDREN:
> Fifth grade is the bomb
> Do that snake
> And sing that song
>
> M-C-C
> Sing that song
> Sing that song
> Sing that song
> Sing that song
> Sing that song
> 204 is number one.

Though poetry may have been MCC's favorite subject, the problem, she related, was that it wasn't necessarily the students' favorite subject:

> They're intimidated by it right away, you know, "How am I going to write a poem?" So sometimes you kind of fill in the blanks to get them started … get them going and pull them out a little bit, and that was what they did with the color thing.

She was talking about the "color poems" that the kids wrote as an exercise in using poetic language. Their assignment was to write a metaphorical poem about a color. Once they wrote their poems, we sat down with small story groups and developed visual treatments for each of them. Once we finished shooting them, I edited them together in the editing lab at the university.

OVER A BLACK FRAME THE VOICE OF A FIFTH-GRADE GIRL,
KATIE, IS HEARD.

> KATIE:
> Hate is black

DISSOLVE TO SHOT OF KATIE SQUIRTING BLACK PAINT ON A WHITE PAPER.

> KATIE:
> It feels like being tugged between two friends.

DISSOLVE TO SHOT OF TWO GIRLS PULLING KATIE'S ARMS IN OPPOSITE DIRECTIONS.

                    KATIE:
               It tastes like rotten radishes.

CUT TO SHOT OF KATIE SPITTING OUT ROTTEN RADISHES.

                    KATIE:
          It sounds like a yell being pointed at me.

CUT TO SHOT OF TEACHER DRAGGING KATIE OUT OF CLASSROOM
FROM BEHIND.

CUT TO SHOT OF KATIE'S HAND SMEARING BLACK PAINT ON
THE PAPER, THEN TEARING THE PAPER IN HALF.

                    KATIE:
                 I hate . . . hate.

DISSOLVE TO SHOT OF KATIE'S FIFTH-GRADE TEACHER PICK-
ING HER UP AND SWINGING HER AROUND HAPPILY, WHILE
KATIE LAUGHS.

FADE TO BLUE.

MCC quickly found that the connection between videomaking and the writing of poetry was truly invigorating the lesson plan and getting the kids more interested in poetry than usual:

> Once they started thinking they were going to write their poem and they were going to videotape their poems, it really just, they just went wild with it ... they really enjoyed what they were writing a whole lot more.

We weren't finished when all "the color poems" were edited. The kids were interested in somehow blending their poems together into one big poetry movie. After some discussion, they developed an overriding theme that centered on the reported existence of a "poetry virus," something affecting the whole school. At the time, computer viruses were not only in the news, but were wreaking havoc on the school computers. They borrowed the virus term to set the stage for their poetry, as if the writers of the poems were inflicted by some sort of mysterious virus—this, as opposed to opening up each poem with a title or explanation of some sort. In this format, the poems could flow more freely because they had a common origin.

It was decided that the assistant teacher, Mrs. Edwards, would read a news-style announcement about the virus. The students loved the idea of using Mrs.

Edwards's voice in their movie. MCC recounted the story of her being cast in the role of the newscaster:

> We had that idea that we were going to do the poetry and I remember because Mrs. Edwards said, "Oh, I'm a terrible reader. I can't believe they want me to read." She's the narrator at the beginning of the film and it's great. It was so wonderful to hear her voice and to remember:

POETRY VIRUS OPENING

<u>CUT TO FAST-PACED COLLECTION OF SHOTS OF CHILDREN INVOLVED IN VARIOUS ACTIVITIES. THE VOICE OF A WOMAN IS HEARD. IT SOUNDS LIKE IT IS COMING FROM A SMALL RADIO.</u>

```
                    WOMAN
              Today in the news
              We have heard word
           Of another virus going around
            Corrigan Elementary School.
           You've heard of the "Love Bug"
               and all the others?
             This is the poetry virus.
             It seems students in 204
                are overcome with
          rhyme, rhythm, and rapture . . .
                    for no . . .
                    for no . . .
                 apparent reason
```

But these young poets caught the poetry bug as they were engaged in thinking poetically. They thought up yet another layer around the virus theme. It had to do with a poem that one of the students had written that had nothing to do with "the color exercise." It was more of a free-form poem by fifth-grader Daniel Duncan, and MCC and the class really liked it. They wanted to find a way to work it into their poetry movie independent of the color poems.

The students thought they could work it in just after the adult voice announced the poetry virus. Visually, they saw it beginning with a fifth-grade boy, played by Daniel, quietly observing the activities around him on the playground. The poem was about animals in a jungle, but since we did not have "animals" per se, and we were engaged in a poetry exercise, it was reasonable to assume we could metaphorically represent the animals of the poem with the "animals" of the playground—

children engaged in various play activities. The children also liked the idea of juxtaposing Daniel's voice poetically with the voice of him as an older, wiser young man later in the poem. The poem then would be about a young man thinking back to his days as a child on the playground and going back in time just by watching the children on the playground of the school where he now taught. It was a very complex idea to say the least.

<u>DISSOLVE TO SHOT PANNING FROM TREES TO A FIFTH-GRADE BOY LOOKING OUT OVER THE PLAYGROUND IN FRONT OF HIM. THE DRUMBEATS AND MUSIC SOUND LIKE MUSIC OF THE JUNGLE.</u>

A MAN'S VOICE IS HEARD AS HE OBSERVES CHILDREN ENGAGED IN ALL FORMS OF PLAY AROUND THE PLAYGROUND. WE LATER WILL SEE THAT IT IS THE VOICE OF THE BOY AS A YOUNG MAN, LATER IN HIS LIFE.

                OLDER VOICE
                Wonders.

THE YOUNG BOY'S VOICE IS NOW HEARD.

                YOUNGER VOICE
  The rocky cliffs lay still in the night's silence

CHILDREN ARE SEEN PLAYING ALL OVER A RAISED PLAYGROUND STRUCTURE. A HAPPY BOY DRESSED IN GREEN SHIRT SPINS AROUND AND AROUND HAPPILY IN THE FOREGROUND.

                YOUNGER VOICE
       The jungle was warm and wet

CAMERA SLOWLY PANS TO KIDS RUNNING ACROSS THE PLAYGROUND FIELD IN SLOW MOTION.

                YOUNGER VOICE
   And the exotic animals made loud noises.

DISSOLVE BACK TO SHOT OF THE BOY LOOKING AT THE PLAYGROUND AND THINKING ABOUT WHAT HE SEES.

FADE TO BLACK.

As with most teachers when I first approached them with the idea of doing some sort of movie in their class, MCC was very interested but didn't feel she had sufficient videomaking expertise. She was worried that she had no idea how to make a movie and she had no idea how to connect it to the fifth-grade curriculum.

When I explained that video had so far tended to work particularly well with challenging aspects of the curriculum, she quickly identified the subject of poetry. She wanted it to be a more engaging and resonant experience for her students. Even though I had found video to complement a wide range of curricular activities, there was something alluring about the *problems and challenges* in learning environments that made sense to apply videomaking activities to. As the adage goes, "If it ain't broke, don't fix it," but if something is broken, why not try a different approach? It made more sense to consider videomaking in light of such challenges and needs largely because of its tendency to energize and activate topics it was associated with.

Videomaking and Poetry

> The poet's job is to put into words those feelings we all have that are so deep, so important, and yet so difficult to name, to tell the truth in such a beautiful way that people cannot live without it.
>
> —Jane Kenyon (Timmerman 2002)

Poetry is actually something video is quite good at. Short-form visual works such as commercials, promos, show openings, and music videos can all be considered forms of visual poetry, especially compared to longer form visual genres like feature films and television shows. Such visual works employ many of the same techniques as poetry even though viewers are not likely to think of them as poetry per se.

Poetry accomplishes its artistry through "incompleteness"; in other words, it places more demands on a reader to actively assemble meaning. The writer of poetry must have a critical distance from the words they use, to see words as a vehicle of an abstract objective, such as a feeling or a complicated idea.

I often use Herbert Zettl's (1990, 326) approach in teaching my filmmaking students poetic visual techniques. The synonym for poetry in a filmmaking sense is the French term "montage." The literal translation is "editing," but in artistic terms, montage is all about poetry—juxtaposing pictures and sounds in such a way that the whole of a message is greater than the sum of its parts. The key is for the filmmaker to invite a viewer's imagination to fill in the gaps between the incomplete images and sounds with meaning. Zettl illustrated this by comparing montage with Haiku poetry: "Haiku poetry often works on the montage principle. In its highly economic structure it implies a larger picture of concept than its lines actually contain" (1990).

Commercials do the very same thing by using words, pictures, and sounds to create abstract impressions of their products. This impression (hopefully) leads to sales when consumers match impressions created by commercials with impressions of what they want to be. For instance, if a colorfully positive visual is juxtaposed with a particular product (soap, beverage, or car), a viewer will tend to intellectually link the two into a combined impression.

If a viewer is involved in making meaning of a visual work, she or he can be said to be interacting with the filmmaker in an abstract dimension. It is this dimension around imagination where new learning opportunities arise, because visually based lessons help learners to see subjects in ways they are not accustomed to in classrooms.

Visual Thinking and Expression

> Visual understanding has intrinsic value, it broadens experience.
>
> —Kevin Barnhurst (1987)

MCC's primary learning objective in her poetry unit was to unlock her students' expressive voices by exercising their use of creative language:

> You try to get them to consider "How else can you say that?" not just like rhyming or putting a word in because it sounds good. "What would be a word that really fits it?" And if you can get them to do that, then they're going to do it in other places and they're going to use more words. Just teaching them vocabulary—it's different when you can say, "Well, what else? Does it just melt? Or does it fade away?"
>
> They feel so pleased in the end when they end up with their piece.

MCC realized quickly into our first videomaking experience that video complemented her broader educational objectives and shared some realizations about the videomaking experience after it was finished:

> For me it really spoke to all different levels. There were kids on that video who read poems that they had written who really struggled with reading—*really* struggled. Sometimes poetry is like that, you know a kid can put words together, and you can help them brush it up, and it's this great, fantastic poem, but they have a really hard time reading something on a fifth-grade level. And it seemed with the video, the kids could really shine. They could see what they wrote, and really experience it, so it really brought the whole group together, and it was fantastic the way that happened.

Abstract concepts are difficult to teach with traditional methods. Video can effectively present and motivate students to apply these concepts in a manner that is relevant to their everyday lives—televisual portrayal. Television is part of their everyday literature.

As Paul Messaris relates, visual messages invite the skull out of its physical constraints to participate in experiences it might not otherwise know: "How is it that pictures, both moving and still, can conjure up a world of almost palpable objects and events despite the many differences between the appearance of the real world and the appearance of any kind of picture, no matter how realistic?" (1994). Visual approaches to education actually open up new ways of thinking and knowing, Messaris continues:

> Proponents of visual education often argue that experience with visual media is not just a route to better visual comprehension but also may lead to a general enhancement of cognitive abilities. To put it differently, the cognitive skills that are brought into play in the interpretation of television and other visual media may be applicable to other intellectual tasks as well. For example, it has been proposed . . . that experience with film and television may improve children's understanding of spatial relationships in the real world. (1994)

Drawing on the work of Olson, Bruner, and Salomon, S. B. Neuman saw the symbol systems of media capable of extending "schemes of thought":

> Not only can a medium implicitly teach a symbol system . . . but it has the capacity to arouse certain general attentional processes and become internalized, serving as a "scheme of thought." While media are complex entities, it is the symbol system employed in each medium that constitutes its most important attribute or essential mode of representation. (Neuman 1991)

In this regard, there are clear parallels between visual thinking and video game dynamics. Marc Prensky (2001) uncovered ten "cognitive traits" in children he described as "the game generation." Borrowing Neuman's term, these traits are listed as *New Schemes* vs. Old Schemes:

1. *Twitch Speed* vs. Conventional Speed
2. *Parallel Processing* vs. Linear Processing
3. *Graphics First* vs. Text First
4. *Random Access* vs. Step-by-Step
5. *Connected* vs. Standalone
6. *Active* vs. Passive

7. *Play* vs. Work
8. *Payoff* vs. Patience
9. *Fantasy* vs. Reality
10. *Technology-as-Friend* vs. Technology-as-Foe

These were clearly distinct ways of thinking compared to previous generations. In highlighting generational differences, Prensky's motive was to call attention to opportunities for fresh approaches to learning.

I recognized many of these traits in K–12 settings I worked in. Over the years of making videos with children I found that "just doing" video with the children seemed to be more natural than teaching the mechanics, process, or critical thinking components of video production. The teachers seemed more curious about such aspects, but the children just wanted to do it, to just get inside of it. Adults wanted to know how everything was done. This has direct linkage to Prensky's "Parallel Processing" and "Random Access" traits. These schemes of thought can be embraced to deliver learning impact where needed.

Visual Thinking in Education

What all this seems to boil down to is that visual thinking is a window into new learning possibilities. A sensible first step is to acknowledge the power and potential in visual thinking, or in the case of K–12 videomaking, "thinking in video."

A second step would be to understand what is already known about visual thinking and expression. This comes down to semiotics in Brown's view:

> Underlying all this visual literacy, media literacy, media education [whatever you want to call it] is semiotics, the study of signs. It studies how signs work and the way we use them; it analyzes how a sign and its meaning are related, and how signs are combined into codes. (1991)

According to Messaris, this requires "greater experience in the workings of visual media coupled with a heightened conscious awareness of those workings" (1994). Barnhurst agrees that there is much more for teachers to know concerning the hows and the whys of visual thinking:

> Just as language is a system of symbols and structures producing meaning, visual artifacts contain a system of space, articulated by formal concepts that combine to convey visual meaning. The definition of "visual literacy" should assert the value of visual knowledge. Visual alphabetization should, on one level, restore the sight of the child to the adult. Instead of seeing only symbols and their hidden meanings, the adult eye might be trained to see again their textures and patterns. (Barnhurst 1987)

The third step, according to all the children I have worked with so far, is to just pick up a camera, jump in, and do it. There is much to discover in the classroom, and video is a spearhead for such discovery.

MCC related one such discovery in *The Poetry Virus*. It had to do with the kids' idea of incorporating an "odd" but meaningful act in a prominent position in their movie:

> Steven . . . he did that twirling thing. That was something that he liked to do. . . . He didn't have to speak and recite a poem, something like that. He could just be in there [making twirling motion with her finger, imitating his favorite act of spinning around and around on the playground] being part of it.
>
> Steven had autism and he was legally blind. . . . He had an odd stance, and he loved spinning . . . and spinning things. And the kids were really good about that. They'd say, "Do you want to spin this?" They can go along with those kinds of things . . . and they admired that he would spin and not get dizzy. He had this graceful way of spinning . . . the kids saw that was something he was great at. Sometimes I think as adults, you think, "Oh, are they making fun?" But they're not. They're like kind of using what works— that's what kids do. . . . They were just able to express themselves in a different way.

Such an opportunity for this "discovery" might not be captured by traditional instructional means.

Living in a Visual World

In the end, thinking in video is about tapping into the visual substance of everyday life. We live visually. We think visually. We make sense of our lives visually.

Not too long ago I found myself engaged in such "visual sense-making" at the wake of a cousin who met an untimely death. Her death happened to be connected to a string of losses that had made the process of closure more difficult. I found myself asking, "Why are so many people I know dying?" It seemed like I was at a funeral every week.

In the midst of my contemplation, I realized I was staring at an easel filled with pictures of my cousin's life. The pictures ranged from shots of her as a little girl to very recent family get togethers. There were close to fifty pictures and they were arranged beautifully, by someone who had an obvious flare for visual presentation. As I looked closer, I realized there were swirling graphics spelling out her name. Pictures of red roses—her favorite—adorning the edges of the poster.

This was a truly professional job. I found out that it was done by a family friend who owned a print shop. He had passed the pictures of my cousin to his "Photoshop Whiz" at the office and asked him if he could put a little pictorial presentation together. It didn't take long and he even ended up designing a funeral program with the visual theme.

Even though it was an uncommonly elegant touch for a funeral, I realized that the process of visually representing life was becoming a mainstay of the ceremony of life closure. Every funeral that I had been to over the past weeks had had a collection of pictures, from rudimentary to sophisticated, but they all had the same effect. They were triggering moving memories in my mind. I was making a movie in my mind of my cousin's life, and it was a fittingly meaningful part of the ceremony of closure. The visual energy of the movie in my mind was adding much to the traditional activities of the wake. It was personalizing my last images of my cousin as I was saying good-bye . . . bringing her whole life into the ceremony, not just the end of it.

It struck me that we are not only living in but also dying in an increasingly visual world. If we live and die visually, we can certainly learn visually.

VIDEOPLAY
Tapping into the K–12 Imagination

> Children do not have morbid fascinations, they have genuine curiosity about the BIG questions and need to process the experiences of their lives in meaningful ways. The big question is, what points of reference are they going to have, who is going to be there for them and what kind of help are they going to get?
>
> —Pat Kipping, 1995a

Movie-Dreaming

I always "movie-dreamed" when I came to Corrigan Elementary, either of movies we had made in the past, like *Coldheart,* or crazy scenes for future movies. A movie-dream is like a daydream except it draws on purposefully imagined experiences rather than "real" experiences. All the moviemakers I know do a lot of movie-dreaming as they gaze at the worlds in front of their cinematic eyes. It's part of the fun of the cinematic imagination.

Laden with the usual gear—camera case, tripod, and utility pack—I had been on my way out of the building after shooting a sequence for a school movie when I found myself walking through the kindergarten hallway. This was a particularly vulnerable area for a movie dream, in fact the place where it all began more than ten years ago.

Room 1 still looked the same. Its most striking feature was its warmth—bright colors, adorably tiny furniture, wall-to-wall letters, numbers, illustrations, and of course, playthings of every shape and form. All the room's features seemed to accentuate the purpose of kindergarten as I understood it—to gently introduce its young citizens to the structures of learning. When I let my imagination go, I could see "video ghosts" of the movies we had made in the room: In the center of the room, I could see Troy passing out the invisible juice; across the quiet play stations, I could see dozens of smiling five-year-olds moving about during the annual pajama party; there was the end of the year graduation song, and I imagined the camera panning the line of little cap-and-gowned graduates.

Room 1's teacher was no less warm than its memories. If Barbara Brighton weren't the Room 1 kindergarten teacher, the next best job for her would be at Disneyworld as the person waving from the balcony of Cinderella's castle.

I had the good fortune of working with her over three school years as she taught each of my three sons and I continued to explore video applications in her kindergarten classroom. Each time I worked with her she became more curious about what I was doing and what we were discovering. The more we did, the more confident she was about making movies herself. In the end, she and the assistant teacher, Sarah DeBella, were the filmmakers and I had become a technical facilitator.

When talking to people about their moviemaking experiences, they referred to them as "videoplays." It added a kind of "kindergarten edge" to what we did. One of the most memorable accomplishments was the videoplay we called *Once Upon a Coldheart*. As far as movie dreams go, this was not all roses—in fact, it was closer to a nightmare!

ONCE UPON A COLDHEART

CREATED BY MRS. BRIGHTON'S KINDERGARTEN CLASS
(A VIDEO EXCERPT.)

<u>INTERIOR—MRS. BRIGHTON'S KINDERGARTEN CLASSROOM—MORNING</u>

RACHEL IS A CHALLENGING LITTLE KINDERGARTENER AS KINDERGARTENERS GO. IT IS ONLY HALFWAY THROUGH THE SCHOOL YEAR AND SHE ALREADY HAS SHATTERED THE ALL-TIME ROOM 1 RECORD FOR THE MOST TIMES A STUDENT'S NAME WAS WRITTEN ON THE BLACKBOARD FOR INAPPROPRIATE BEHAVIOR.

THAT'S WHY IT IS NO SURPRISE THAT WHILE THE REST OF THE CLASS IS OBEDIENTLY RECITING THE ALPHABET TO THE

BEAT OF MRS. BRIGHTON'S POINTER, RACHEL IS INSTEAD PERFORMING A MEDLEY OF GYMNASTIC EXERCISES IN THE BACK OF THE CLASSROOM. THE FUN FOR RACHEL IS OVER BY THE LETTER *K*.

                  MRS. BRIGHTON
                Excuse us, Rachel.

INTERRUPTING IN A SWEET SCOLDING VOICE SHE COULD ONLY HAVE LEARNED IN KINDERGARTEN TEACHING SCHOOL.

                  MRS. BRIGHTON
   I'm afraid that if you don't join us at the front of
   the room, I will have to ask you to sit at your desk
              with your hands folded.

RACHEL RESIGNS HERSELF TO THE SCOLDING AND HEADS OVER TO HER DESK.

                  RACHEL
                  Hmmphhh!

RACHEL GLARES, THEN BURIES HER FOREHEAD ON HER SOFT ARM. WHILE THE CLASS CONTINUES THEIR ALPHABETIC RECITATION . . .

                  CLASS
            "L . . . M . . . N . . ."

. . . EVERYTHING SEEMS TO SLOW DOWN.

                  CLASS
     "O . . . O l i v i a   C o l d h e a r t !"

RACHEL RAISES HER HEAD FROM HER DESK.

                  RACHEL
              Olivia Coldheart?

A FRIGHTENING, DARK-HAIRED WOMAN IN A FLOWING BLACK DRESS IS STANDING AT THE FRONT OF THE ROOM BESIDE MRS. BRIGHTON.

                         MRS. BRIGHTON
            I can call her Olivia because she's my sister. You
            can call her Ms. Coldheart. While I'm away for the
            next few days, my sister will be taking my place,
              and I expect all of you to be on your very best
                                behavior.

MRS. BRIGHTON GATHERS HER THINGS AND WALKS TOWARD THE
DOOR. WHEN SHE EXITS, ALL HEADS TURN IN UNCERTAIN UNI-
SON TO THE DARK FIGURE, NOW ALONE, AT THE FRONT OF THE
CLASSROOM. SHE SMILES.

                         MS. COLDHEART
                            This . . .

SHE HESITATES FOR DRAMATIC EFFECT AND THEN, JUST AS
QUICKLY AS IT HAD APPEARED, THE SMILE VANISHES INTO
THE DARKNESS AROUND HER.

                         MS. COLDHEART
                 . . . is Olivia Coldheart's class now!
              Forget about that Mrs. Brighton—she's gone.

THE CHILDREN GASP LOUDLY.

                         MS. COLDHEART
                 There'll be no more reading stories.
                        No more free choices.

LITTLE FIONA'S MOUTH WAS FROZEN WIDE OPEN.

                         MS. COLDHEART
                   Don't think that we'll be on the
                      playground anymore, either!

THE CHILDREN ARE IN SHOCK. THIS CANNOT BE HAPPENING!
BUT IT WAS, RIGHT IN FRONT OF THEIR FIVE-YEAR-OLD
EYES.

                         MS. COLDHEART
               And those little snacks you look so forward
                  to in the afternoon—we won't be having
                   snacks anymore. So don't be hungry!

ALL THE CHILDREN JUMPED BACK. HOW COULD SOMEONE BE *SO* EVIL AND SMILE THE WHOLE TIME?

> MS. COLDHEART
> There just won't be any more fun in this room. And that *especially* goes for you, Rachel!

ON THAT NOTE, MS. COLDHEART POINTS HER FINGER STRAIGHT AT RACHEL. HER NAME SEEMS TO ECHO THROUGH THE WHOLE ROOM.

> MS. COLDHEART
> Rachel! Rachel!

THEN SOMETHING EVEN STRANGER HAPPENS. ALL OF A SUDDEN, MRS. BRIGHTON IS BEHIND HER, GENTLY SHAKING HER SHOULDERS.

> MRS. BRIGHTON
> Rachel. Rachel, sweetie. It's me!

RACHEL TURNED BACK TO MEET MRS. BRIGHTON'S WARM SMILE.

> MRS. BRIGHTON
> I think you were having a bad dream.

RACHEL SNAPS OUT OF HER GROGGY STATE.

> RACHEL
> Huh?

SHE HAD INDEED FALLEN ASLEEP SHORTLY AFTER SHE HAD PUT HER HEAD DOWN. THAT MEANT OLIVIA COLDHEART WAS JUST A DREAM. RACHEL FEELS A TREMENDOUS WAVE OF RELIEF, AND THEN REGRET—REGRET FOR HER BAD BEHAVIOR THAT HAD STARTED THIS WHOLE INCIDENT. SHE REALIZES HOW WRONG SHE HAD BEEN AND HOW GOOD SHE HAS IT WITH MRS. BRIGHTON AS HER TEACHER.

> RACHEL
> Mrs. Brighton, I promise I'll never be bad again!

RACHEL FLINGS HER ARMS AROUND HER TEACHER.

                       MRS. BRIGHTON
Oh Rachel, that's so nice. I knew you would come around. Now that you're ready to be a good girl, I want you to come over here and meet someone special. It's my sister, Olivia!

RACHEL'S MOUTH FALLS OPEN IN HORROR. IT COULDN'T BE . . . COULD IT? PICTURE FREEZES ON RACHEL'S WIDE-EYED EXPRESSION, THEN FADES SLOWLY TO BLACK.

The Making of *Coldheart*

Once Upon a Coldheart was a "second-generation" kindergarten movie made with Mrs. Brighton's class. The first generation of kindergarten movie was made *before* I realized just how visually aware kindergarteners were.

The second generation was about giving children a chance to use video to say or do something *they* wanted to do. We were going to let the children make *their* video plays, rather than *mine*. In my first generation of videomaking (Grades K–2), I had slowly realized I was approaching the moviemaking process in too didactic a manner. This time, I was going to see what the kindergarteners could do if I gave them more creative license—to see what they were capable of.

Mrs. Brighton and I started out by talking with the children about what they wanted the movie to be about. The children leaned heavily toward some sort of fantasy or make-believe story with a scary character in it. There was no doubt, these kids wanted to play with imagination.

> i·mag·i·na·tion, n
> the ability to form images and ideas in the mind, especially of things never seen or never experienced directly. (*Encarta World English Dictionary* 1999)

Albert Einstein once said that music was a driving force behind his theory of relativity. His scientific discovery was a result, he explained, of intuition driven by musical perception. What this implies is that the thought process transcends mere use and consideration of raw ideas. In Einstein's case, music was a garden for the cultivation of creative thought.

This said, music certainly does have a way of igniting thoughts and feelings. When I coach my TV and film production students on the use of music in their stories, I remind them that music tends to bypass the sensory guards and filters that organize our everyday lives. If they want viewers to "feel" their stories, they can use music as a "live wire" to trigger viewers' emotions. For instance, if they want us to feel happy about a scene, playing happy music will make viewers read the scene in a happy way, without a level of rational consideration. If they want us to feel a character's sadness, they can play sad music and it will encourage us to take on that perspective.

Music in this sense, according to Edwin Gordon, a recognized authority on the psychology of music, has a way of igniting imagination. Gordon wrote, "Through music a child gains insights into herself, into others, and into life itself. Perhaps most important, she is better able to develop and sustain her imagination. Without music, life would be bleak" (1990).

In my K–12 videomaking experiences, I have found that video seems to have a transcendent effect not unlike music on people who watch it, and this can be used to the advantage of learning. Moving-image stories trigger imagination in a similar way as music—they create an expanded arena of perception that can stimulate ideas and foster intellectual growth. This was evident throughout the creation and viewing phases of the *Coldheart* movie.

Mrs. Brighton and I were, in effect, playing with the children in a playground of imagination. Many times when I was working with them, I felt like they were taking me on a figurative tour of their "Neverland." This was a place where there were no rules, no barriers, and no adult structures in place. It was a place where they could play out all kinds of scenarios that real life did not always provide for them. But like the toys in Room 1, this playground was not all play. There was always a learning task lurking nearby that could be attached to play activities if teachers wanted to bring it in.

The *Coldheart* Story

As the children and Mrs. Brighton and I were developing the movie story for *Once Upon a Coldheart*, it was Rachel who brought in an idea from a story she had read with her mother called *Miss Nelson Is Missing* (Allard 1977). In the story, a teacher has some problems with her students' disruptive behavior and decides to teach them a lesson. She comes to school disguised as a very mean substitute teacher and ultimately makes her students appreciate how good they had it with her. Once they learn their lesson, Miss Nelson returns to the class and they behave exceptionally well. They know the evil substitute could always return if they were to act otherwise.

After a lot of discussion, we outlined a story that was somewhat similar in terms of its "Golden Rule" underpinning but very different in terms of its plot. Instead of an entire class misbehaving, we decided, for the sake of order, to focus on one child misbehaving. Rachel was chosen to play this character partly because she was naturally outgoing, and it was reasonable to assume she would be comfortable acting in front of the camera. In real life she also happened to be the opposite of the character she would play. Having a well-behaved child acting out misbehavior would not encourage excessive celebration of such behavior.

Instead of a scary mask, the teacher in our story would have, with the aid of video tricks, an evil twin. We had done a video twins exercise some years before so this was not a new trick, at least to Mrs. Brighton, Mrs. DeBella, and me. In the end, the trick became the highlight of the movie.

We shot everything but the Coldheart sequences first: Rachel's antics, the obedient class reciting the alphabet, close-up shots of the children reacting in fear to the evil Ms. Coldheart (even though she wasn't there at that time), and all of Mrs. Brighton's lines.

The acting and pretending aspects of what we did seemed to serve two functions. First, they introduced the concept of stories as manufactured narratives. We talked about what we wanted to happen in our story and then we acted it out. The acting was, for all intents and purposes, an exercise in "faking it." The children were asked to pretend they were scared at a time and in a place where there was nothing to be scared of. The pictures would then be used to "trick" viewers into thinking they were actually scared. They delighted in this trickery.

Mrs. Beechwood shared a similar reaction to the experience of acting when it came to her first-grade movie. There was a scene where she had to pretend that she was missing her first-grade students after they had just gone home for the summer. In reality, the children were watching and listening behind me while I was shooting the scene, openly coaching their teacher to "act sad." Mrs. Beechwood later related,

> We were showing it to all the parents, and little Harry said, "She's just pretending she misses us." He was talking about when we did the movie—I was pretending—but it came out to the parents, it sounded like "She doesn't really care about us, she's just pretending she's going to miss us." It was so classic . . . out of the mouth of a six-year-old. They were just perfect.

The other thing that acting for the screen seemed to do was to give children an alternative space to act out things they might not be able to in everyday life. Rachel's antics, for instance, allowed for the disruption of a class activity. Students involved in the scene had the opportunity to experience something their normal lives would not allow to happen without harsh reprimand. In a sense, the acting allowed for behavioral catharsis without penalty.

The exercise introduced and solidified the notion that narratives are purposeful illusions of reality. This could be a helpful experience in making them healthy skeptics of stories manufactured by other storytellers.

However, as we were doing it, it did raise the question, "Are five-year-old children better off knowing the truth about film and television stories, or is it better to leave the illusions under wraps for a later, more developmentally appropriate time in life?" Based on our moviemaking experiences with past kindergartners, there was no compelling reason *not* to give it a try. After all, we would be building a foundation of media education based upon truth.

There was much anticipation when it came time to shoot the Coldheart scene. Mrs. Brighton changed into the Coldheart costume behind closed doors. When she returned in costume, it was very difficult for her to perform as the mean character without laughing. Several of the kids were actually coaching her into the evil role. Most of the children seemed genuinely energized by the make-believe antics and, overall, naturally comfortable in their role as young filmmakers and actors behind the visual conspiracy.

This more active, second-generation media lesson was clearly paying off. The children were much more engaged in "their" exercise than they had been in "mine," mostly because they were coauthors. This was videoplay anchored in their imaginations and they were "playing" at abandon. They were also thinking, interpreting, and solving problems while they were playing.

For instance, when we decided not to have everyone in the class misbehave, and it would be just the character of Rachel, we discussed the "inappropriate behavior" she should engage in as co-conspirators. This was putting them in the position of evaluating their own behavioral limits. In the end, they were very careful not to make Rachel be "too bad" or disrespectful. They did this on their own.

They seemed delighted in the fantasy of it all, being architects of an imaginative world of make-believe. They were very interested in "playing with" ideas they knew they should not directly experience, like bad behavior, evil kindergarten teachers, and sharing someone else's nightmare. They placed themselves squarely on the line between real and make-believe and were very comfortable there.

The reassuring part of the "scary" *Coldheart* story was that the experience was occurring in the supportive and nurturing environment of a kindergarten classroom. When given the opportunity, the children were facing their fears by coming up with a list of things the evil substitute teacher could take away from their positive classroom environment—snack, playtime, and so on. They also wanted the story to be truly and believably scary to those who watched it. In fact, it seemed the most engaging part of the creative exercise was in developing the evil character. This seemed to be at the center of their creative efforts and the strongest motive behind their storytelling process: They almost seemed to want to scare themselves. In the end, they actually did!

A Teacher's Perspective on *Coldheart*

Some months after we had made the *Coldheart* movie, Mrs. Brighton shared some perspectives on the experience—in particular, how her students walked the line between reality and fantasy.

> When we were filming *Coldheart* it was fun for me because I remember getting into character—it was the side of Mrs. Brighton that's never emerged.
>
> We were filming and I was being Mrs. Brighton, and all of a sudden it was time to be Ms. Coldheart. The kids were all sitting together and they saw me carry the dress and wig into the bathroom, and I came out with it on and at least half of them were terrified. Where was Mrs. Brighton? They could not figure out where Mrs. Brighton was and it was just so fun!
>
> It's really learning to understand reality, and separating what you see on video and understanding that video allows you to make lots of things look very, very real. For some kids, the concepts are not going to be quite in place developmentally. I think for a lot of the kids it's so good to start to separate, because I think we're in an age where kids need to know that what they're seeing on television is not real.

The most elaborate illusion we created was Ms. Coldheart herself. The most crucial moment in this scene was when both Mrs. Brighton and Ms. Coldheart were to be shown in the same room together. If we could successfully make a viewer believe that they were in the same space at the same time—even though this was impossible since they were the same person—Ms. Coldheart would come to life as a character.

This is an example of *suspension of disbelief*: a term to describe the depth of involvement a viewer is willing to give to a fictional story. When they sit down to watch a movie, audience members go through a preliminary stage of story engagement.

First, they take in basic story information and consider whether the story is worth experiencing. Even though they have made a preliminary decision to participate in the story, there still is time to gather information on whether it is the kind of story they might find interesting. They can always walk out of the movie theater or change the channel, depending on how they feel about the story.

At about the same time, viewers eye the authenticity of a fiction story. They ask, "Is it believable?" or "Is it a complete fabrication?" Believable stories are preferred as a rule because we usually prefer to surrender our imaginations to the story and let it take us away as viewers.

Storytellers must therefore prepare their stories for such viewer demands. This usually means we have to get to the point of the story quickly, usually by beginning with some sort of action. Action encourages participation in the story, because it moves like our lives. And our stories and characters have to be believable and authentic if viewers are going to relate to them as if they were real people.

If viewers are sufficiently engaged in a story as they seek information and authenticity in it, then we can expect that they will suspend their disbelief in the artificial act of, among other things, watching a story unfold on a clearly artificial screen, and they will join in the story experience as if it were an event in their lives.

This is almost a form of self-hypnosis story audiences go through so that they can leave behind the complicated real-life ambience of movie theaters and living rooms surrounding them as they watch, and just enjoy the stories in front of them. Though a fairly complex story process, it is also fairly innate, even in young children.

Reading books also requires a degree of suspension of disbelief. A good book will make readers forget about the environment they're actually reading in and transport them to imaginative lands of their stories. Reading stories in print doesn't require the same kind of suspension of disbelief as a video story. This is because words trigger the mind's eye to illustrate for them. Many reading enthusiasts cite this as a reason why reading book stories is better for children than "reading" moving-image stories. "All the imagination is taken away from the reader," they say.

Even though videomakers do offer more visual cues for their readers, there is no less need for imagination. In short, videos are only pieces of picture and sound information that must be linked together to form a whole idea that is greater than the sum of its parts. This is just a different way of using reader imaginations.

S. B. Neuman shared some insights in this direction. She reduced the discussion to the concept of symbol systems:

> That television's unique symbol systems might influence the way children process information clearly hit a responsive chord among a number of media researchers (Clark and Salomon, 1986). No longer was the content as important as the symbol system used to convey it. Television, for example, calls upon people's understanding of production techniques, changes in visual perspectives, action and music to interpret its messages. . . . His research [citing Salomon's experiments in 1974 and 1979] indicates that TV programs, making use of specific cinematic techniques, may facilitate the acquisition of certain cognitive skills as attention focusing and perspective-taking (1974). (1991)

In the case of these young videomakers, we were involved in the other side of the perspective business—perspective giving.

Video has a relatively high suspension of disbelief factor, meaning that we have to pay more attention to authenticity and devise clever illusions of reality because life through a video camera looks a lot like everyday life. The closer to real life a media form looks, the more demanding it is to uphold the suspension of disbelief. A theatrical performance on a stage has a much more tolerant suspension of disbelief and therefore does not require the same attention to detail that film stories do. If a woman dances across a stage singing "I am the wind," we might give her the figurative benefit of doubt since theater often involves a higher level of abstract expression. We don't have the same luxury in a movie.

This is why we didn't just say that Mrs. Brighton had an evil twin. And it might not be enough to just see the twins in completely separate shots. We had to show the evil twin in the same space with Mrs. Brighton so that viewers would "believe" they were actually twins.

First we shot a close-up shot of Mrs. Brighton sitting down on a chair at the front of the classroom, talking to the kids about her sister. We followed this up with reactions of some children looking at her while she talked. Then we went back to Mrs. Brighton on a slightly longer shot to show that someone in a dark dress was standing next to her. Only the torso of this person was visible. It was actually Mrs. DeBella, the assistant teacher of the class. She was being used as a "body double" for Ms. Coldheart. Then as Mrs. Brighton looked up at her sister, we followed with a shot of several kids shifting from looking at Mrs. Brighton in a chair, to the standing figure above her eye level. This was followed by the payoff shot, camera tilting up like the children's eyes to reveal Ms. Coldheart.

Because of another pair of cinematic phenomena—"persistence of vision" and "montage"—the illusion of Ms. Coldheart was complete with this lineup of shots. This is because viewers inductively create meaning out of individual shots by blending and relating them as they come up on the screen.

Mrs. Brighton and I were truly exploring the line between *the fantastic* and *the real* with these young videomakers, and promoting the idea that imagination was a fun place to be as both consumers and authors of stories.

Watching the *Coldheart* movie

Something very interesting and unexpected happened when the children sat down to watch their completed movie: Some of them actually believed it! Mrs. Brighton explained:

> When we sat down to screen it, some of the children actually wondered whether the story was true even though they had witnessed it. "Was that really you?" one asked. I actually believe there were some kids who graduated from kindergarten that year who thought Ms. Coldheart might be real.

Mrs. Brighton saw a parallel between this unexpected finding and a kindergarten math concept called "conservation."

> Here are five pennies. If I close my hand and hold them over here, how many are in there now? . . . Some of them don't have that concept. But it's going to be exactly the same because I haven't taken one away. Conservation

would be that if you have any number of objects; they could count and were able to conceptualize that number—seven or three or whatever—and then you cover it up or even put it behind your back, is the number the same? Or you can switch them to the other hand and say, "How many are in my hand now?" and they may not know that it's the same. They will need to count them again. But I think it's similar [with viewers] . . . and when they're ready, it occurs.

But as I compare it, I think of the very similar thing, put a dress and a wig on Mrs. Brighton and they wonder if she's the same. But I wonder if it correlates in their development . . . if those would be the same kids who couldn't do that.

. . . It's the kind of thing that will come with experience and, I guess, we do work on it because we want to provide them with experiences, but it's just like learning a motor function . . . you can't make a child walk until he's ready.

I still think there were kids that could never quite figure out if that was me or not. Mrs. DeBella was by Mioshi, who kept saying, "That's not Mrs. Brighton, that's not Mrs. Brighton, that's not Mrs. Brighton," and you wonder . . . if she needed to convince herself that Mrs. Brighton couldn't get up there and be that mean, or that the wig and the dress transformed me enough, that she was either talking herself into or out of something? But she was reciting this little idea underneath and I thought that was really interesting . . . but I think for some of them it was kind of real. Even the next day I recall some questions like, "Was that really you?"

Imagination is like an expansion chamber for the classroom—a virtual learning environment reserved for everything from difficult issues to pure play—just a different "venue," as MCC would say.

A Venue for Dealing with Tough Issues in Life

Teachers are ideally situated to help kids make sense of the real world and even of difficult issues: everything from death and dying, current events, and TV and movie stories. Pat Kipping elaborated on this idea, suggesting that we need to give children more credit for their ability to handle everything from "the truth" to very complex ideas. If we don't, others will.

Maybe because we, as a society, are so idealistic about children and their innocence, we have ignored their genuine curiosity about the BIG questions about life and death. In ignoring it, we have created a vacuum that is being filled, not be people genuinely interested in our children's healthy development or in sharing their own insights but smart enough to identify a need and fill it to meet their own needs—profit. (1995b)

But it's not simply delineating the line between real and fantastic. It's also about learning to appreciate and celebrate the worlds of each. Videomaking allows us to get inside the playground of imagination where we can stand in each others' shoes, share feelings, share understandings, share differences, and get to know ourselves and each other better. As Mrs. Brighton related, video was a doorway for new experiences and lessons:

> They loved it! They wanted to see it again except some were scared of it. Ms. Coldheart was so convincing and effective that they were a little frightened. It made it easy because I could always say, "You know, Ms. Coldheart might stop by tomorrow. I think we should be good."

CHAPTER 11

INVENTIVE VIDEO

> Our schools must do more than convey a set of facts or even a set of basic skills. They must, above all, teach young people how to think. The basic process of thinking—critical thinking, analytical thinking, disciplined thinking—is at the heart of every other skill students learn in appreciating a work of literature, to writing a persuasive essay.
>
> —Anthony T. Podesta, founding president, People for the American Way (cited in West 1997)

EVERYTHING I would expect in a K–12 videomaking experience was unfolding in Joseph Hamilton Mann's class—"the expected" and a whole lot more. In line with my expectations, they were learning a lot about media production and programming and becoming increasingly media savvy. They were actively utilizing their imaginations, which was making their course content more engaging to them. They were exercising their voice and expressive spirit so that issues in the class were more relevant to them than mere words memorized from a blackboard. They were socially bonding in ways Mr. Mann had never seen before, and this was truly significant given the racial divide in the class.

"Mann," as he was affectionately referred to by his students, was teaching two sections of a Business Law class for grades 9–12. There were fifty-five total students in the two sections, all but fifteen African American.

"The most amazing thing to me as a teacher," Mann reflected after the experience, "is that most diverse classes like this are taken over by the white honor students.

In this class, the leaders were all black, every one. And nearly every one of them was having difficulties in other classes."

However amazing these dynamics were to Mann, they were typical factors of the K–12 videomaking experiences I had experienced up to this point. But there was one aspect of Mann's classroom experience that truly stood out: the "inventive video" factor. This is a term I borrowed from a primary school learning technique called inventive spelling.

Inventive Spelling

It was first-grade teacher Sally Beechwood who first introduced me to the term. I must admit, I was skeptical when she explained it to me. After all, what if it didn't work? What if it just engineered a generation of poor spellers—people with no regard for discipline and consistency in written communication?

The idea behind inventive spelling (Adams 1990) is to let a young learner's desire to write motivate their adherence to proper writing form and procedure, rather than vice versa (*Using Inventive Spelling* 1999). Traditional spelling instruction has tended to emphasize form and correctness *over and before* expression. In other words, students are encouraged to write correctly before they are invited to let their thoughts flow in writing.

Mrs. Beechwood agreed with the inventive approach to spelling. To her, the traditional method could discourage young writers from expressing themselves and could therefore instill a negative connotation around writing.

Expression and form were simultaneous and synergetic learning outcomes with inventive spelling. There was no proper order to learning them. Demanding "correctness" before writing could stifle young writers.

Write First, Spell Later

Inventive spelling allows students the opportunity to freely express themselves with words. Teachers accomplish this by legitimizing the words their young writers write as meaningful expressions. The objective is to reward spontaneous expression and develop a sense of comfort and confidence in writing. Spelling and form should not create barriers to expression.

This is not to say that form and correctness are not important. In fact, proponents (Adams 1990) have demonstrated the technique motivates an interest and curiosity in correctness and word recognition that makes them better spellers in the long run. This is because their writing drives form, not vice versa. Spelling is then seen as a tool of more effective expression. Students also gain an appreciation of form and protocol through their reading experiences. Teachers can model spelling and writing strategies by calling attention to patterns in works they read as much as in works they write.

To Mrs. Beechwood, this technique unlocked the *writer within* and created a critical thinking framework for them. Critical thinking was important to her because she was in a position to build foundations of her students' thought processes.

Mrs. Beechwood was a firm believer in teaching critical thinking and, as Anthony Podesta (West 1997) would characterize it, teaching them how to think on their own.

Critical Thinking

> The ideal critical thinker is habitually inquisitive, well-informed, trustful of reason, openminded, flexible, fair-minded in evaluation, honest in facing personal biases, prudent in making judgments, willing to reconsider, clear about issues, orderly in complex matters, diligent in seeking relevant information, reasonable in the selection of criteria, focused in inquiry, and persistent in seeking results which are as precise as the subject and the circumstances of inquiry permit. Thus, educating good critical thinkers means working toward this ideal.
>
> —Peter A. Facione, Dean of the College of Arts and Sciences, Santa Clara University California (*Critical Thinking* 2004)

As a teacher, I have found that critical thinking is difficult enough to define, let alone teach. It's the kind of idea that usually takes at least a few sentences to explain, and if it takes any less than that, the words overheat. Take this definition of critical thinking, for instance: "The art of thinking about your thinking while you are thinking in order to make your thinking better: more clear, more accurate, or more defensible" (Paul, Binker, Adamson, and Martin 1989).

But when critical thinking actually happens, it is tremendously rewarding because it marks the point at which education "takes flight" and becomes the learner's lifelong possession to use for living. Critical thinking in everyday practice occurs almost unconsciously, not unlike the process of decoding letters, which occurs with an agile reader. Those who think critically don't stop to realize, "Hey, I'm thinking critically." They just do it and it feels like any other thought process they engage in—invisible.

The teachers I know delight in critical thinking because it spreads the spirit of learning outside the classroom, into everyday life. But teachers aren't the only ones who see its value. Business leaders and educators in the Partnership for 21st Century Skills observed,

> Today's education system faces irrelevance unless we bridge the gap between how students live and how they learn. Schools are struggling to keep pace with the astonishing rate of change in students' lives outside of school.... People need to know more than core subjects. They need to know how to use their knowledge and skills—by thinking critically, applying knowledge to new situations, analyzing information, comprehending new ideas, communicating,

collaborating, solving problems, making decisions.... [They] need to become lifelong learners, updating their knowledge and skills continually and independently. (Thoman and Jolls 2004)

Another business group, the Business Coalition for Education Reform, published *Why Business Cares about Education,* which highlighted changes in expectations of future employees and their educational backgrounds: "In order to survive, businesses need individuals who possess a wide range of high-level skills and abilities, such as critical thinking, problem solving, teamwork and decision making" (Business Coalition 2001). Vast and continual changes in the economy and business landscapes over the past fifty years had created an environment that was now "fueled by brains rather than brawn." The coalition looked to educators to respond to this evolving business need.

Videomaking and Critical Thinking

It was Mrs. Beechwood who first helped me see a connection between critical thinking on a media literacy level and critical thinking on a fundamental learning level.

I had always seen our videomaking experiences as a means to get children to develop a critical perspective of media. We were, after all, "playing around" with media tools and constructing media messages. The first-graders, their teacher, and I were using media to engage a public in our ideas. The children were therefore in a position to see the media-making process in a different way than they were accustomed to: from a creator's rather than a consumer's standpoint. They were now "behind the curtain" that had once concealed media-making activities from them. Making video was encouraging them to question the media that they watched from a different perspective—from the perspective of storytellers engaged in the process of communicating a message.

One of our most memorable experiences together was the time we visited a television studio and showed them how "to fly." We set up a studio scene with a blue curtain background and then put a table covered with a blue fabric in front of it. Then we had one of the students, Zoe, lie on the table (face down) and pretend she was flying like a superhero. Using a video effect called chroma key, we substituted the blue-colored elements in the scene with footage of an open blue sky, creating the illusion of Zoe flying through the air.

Not only did the children love watching the magic of the special effect, they also left with a more informed critical perspective of media. They had witnessed the trick behind a video illusion that they could apply to media stories they watched. We had promoted a notion that TV consists of carefully manufactured (not "real") pictures and sounds. We also spent time talking about other video illusions and effects, such as camera angles, lighting, picture-sound combinations, and animation. They were being sensitized to the illusions behind the media in hopes that they might develop a healthy skepticism about what they watched on television.

This health skepticism was something that could actually make television better in Roger Silverstone's view, something he called democratic literacy.

> This relates to the whole concept of a critical conscience among the citizenry, not mesmerized by a medium that can distort reality and social priorities, but rather discerning and demanding of that medium and, through it, challenging the very structure of society. This suggests Freire's thesis that, in the broader sense of communication, all people are to participate in building their national culture. (1994)

Mrs. Beechwood reflected on what the children walked away with from the experience:

> They got a little bit of the sense that . . . production is real. Stuff just doesn't happen. . . . When something unusual happens and you say, "Why do you think that happened?"
>
> They'll say, "I don't know." They really never give it a thought. Their thought processes doesn't go beyond that. . . . The movie showed them that. . . . We put it on film and when they do it themselves, there's more to it . . . especially when we did the little girl flying. I think they just take all that in and a lot of them at face value. I think maybe that showed them there's more to it.

Fundamental Learning and Critical Thinking

The critical thinking going on around our videomaking experience transcended issues of video and media. Sally explained this after we had finished our second movie together.

> When you're teaching kids in a setting where the economic structure and the home structure and the skill level are so varied, some children—their world is very small. And I think anytime you can expand that with outside experiences you're expanding everything about them. You're giving them a new reference point, you're giving them new knowledge, you're giving them more critical thinking . . . new experiences they can relate to and I think it's really increasing their brain power.

There was something about putting a video together with a group of kids focused on the same task that naturally inspired a critical thought process, especially when applied to a learning objective of some sort. Perhaps it was the fact that we were deliberately preparing a public presentation and thinking about how it would be received by them: a self-reflexivity of sorts.

David Gauntlett found a similar critical spirit in his videomaking activities with children at the Beckett Park School in Britain. They seemed to enjoy having the opportunity to make links between their views and the local area and to examine inconsistencies between their groupmates' professed beliefs and actual behaviour. . . . [Their teacher] was consistently impressed by the quality of their video work, and indeed was surprised to hear of specific pupils doing particularly well—producing interesting arguments or novel ideas—when their written work was generally of a lower standard. (1996)

For whatever the reason, the process of creating a video was consistently cultivating critical thinking, and Joe Mann's class was no exception when it came to this phenomenon.

Videomaking and Business Law?

I was skeptical about how Mann was going to make a videomaking connection to the subject of Business Law, let alone how he might get his students to think critically about it.

The purpose of the course, as Mann saw it, was to educate students on the various aspects of law that converge in communities around them. He had set out to demonstrate how civil, criminal, federal, state, and municipal laws interacted with each other and how understanding relationships between them could help communities solve problems.

Mann had been talking to my wife, Sharon, and me about the videos we had made in K–12 settings, and expressed great enthusiasm about trying it in his law class. Sharon volunteered to facilitate, not only to help a friend but also because so many in the class were at-risk students—at risk of dropping out of school. She believed they would be more likely to stay in school if they were part of a video project. She had already worked with kids from nearby middle schools and successfully applied videomaking to their after-school program.

Midland-King High School was a predominantly African American high school adjacent to the city's troubled south side. Midland-King's dropout rate was disturbingly high, something that school administrators "preferred not to talk about," according to Mann. There was also a very high crime rate in the area, with regular news reports about poverty, unemployment, drugs, guns, and violence.

Sharon brought a video camera with her on her first visit to Mann's class and got the kids shooting right away. There happened to be a special guest in the class that day speaking about an issue of federal law that had just come to light in the area: the Racketeer Influenced Corrupt Organizations Act, otherwise known as RICO.

City housing official and community leader Michael Mohammed explained that RICO was not only relevant to the subject of law but also to the lives of each of the students. Passed by Congress in 1970, RICO was, over the first decade of its existence, used to destroy the Mafia. However, in the 1980s, the act began to spill beyond the Mafia into areas well beyond its original purpose. One of these areas was the prosecution of gang-related activities.

The subject of *gangs* was a real sore spot in the community around Midland-King. Violence among rival gangs in the area had wreaked havoc on the school and its surrounding neighborhoods and families, and the problem seemed to be getting worse over the past year. Nearly all of the students in Mann's two classes had been touched in some way by gang violence or had known young people who either were in gangs or were affected by gang violence, including murder.

Mohammed's talk about RICO set the stage for the class, their video project, and a lasting critical perspective on law they would not soon forget. After Mohammed left, the students agreed they had to make a video about the effects of RICO on their community. It was clear that this video was tapping into much more than a class project, but the class project had unearthed a vital ingredient to a successful K–12 videomaking experience: purpose.

Anyone can turn on a video camera and shoot pictures of this, that, and the other thing. It's an entirely different experience to connect moviemaking to a purpose. Purpose frames a video like a thesis frames a paper. It makes it readable and relevant to a public outside of its making. It amplifies content so that it can be "heard" by a public outside of its making. Purpose also unlocks the moviemaking mind.

Unlocking the Moviemaking Mind

Any TV or film production student I have ever worked with has at least a thousand movies in his or her mind. These mind's-eye gems come in all shapes and sizes: pictures of the past, dreams of the future, quirky eccentricities, and even a few great screenplays that would rock the box office. The movies in their minds even have special effects: fast-forward, reverse, slow and fast motion, freeze frames, wild juxtapositions. And they're much easier to make than movies they watch at the theater: They just have to close their eyes and open the curtain of their imaginations.

One thing that seems to unite the moviemakers I work with is the immense glory we experience in sharing our stories with other people. I often tell them it's something we moviemakers share in our moviemaking DNA. Sharing our visions gives us purpose. But within this need to share lies a significant challenge to moviemakers: having something to say. The actual work and craftsmanship of moviemaking (though backbreaking as a rule) is much easier in comparison.

I often tell my students that a moviemaker is the metaphorical equivalent of a spotlit figure standing on stage in front of a large audience. All eyes are upon the figure. What does he or she do now? This is their greatest challenge. Hopefully, it will be a worthwhile and fulfilling experience for the audience. The best we can hope for, as moviemakers, is that our audience will leave feeling moved by our works. My students often talk about their dreams of making someone else feel as good as they felt after watching a great movie. They want to give back.

Having something to say and truly "giving back" are not arbitrary acts of pointing the video camera at whatever is in front of us. These aspirations involve connecting the camera to the purpose within us and articulating it aurally and visually for others to appreciate. The value in this for others is in experiencing perspectives from points of view other than their own. As writer Gary Provost advises in his book, *Make Your Words Work,* "I can feel like me anytime" (1990). When we watch stories, he added, we delight in standing in the shoes of people other than us.

But in seeking this "otherness," audiences don't expect to watch and listen to things they already know. Michael Rabiger advises those who want to make films "to conceive works that participate in modern thought and modern dilemmas, and that prompt questions and ideas cutting across conventional thinking" (2003).

This is also where critical thinking comes into play, because thinking critically is, in a sense, prompting questions and ideas that cut across conventional thinking. Putting a video together involves a form of thinking out loud about an idea that is not unlike critically examining it. In this light, Paul, Binker, Adamson, and Martin's definition of critical thinking comes into clearer focus: "The art of thinking about your thinking while you are thinking in order to make your thinking better: more clear, more accurate, or more defensible" (1989).

This thought process goes hand-in-hand with the process of making a film or video, largely because it is a work that is being made to positively affect others. It is like translating the tumultuous flow inside our creative and critical thought process into a clear and articulate work others can understand and, hopefully, be moved by.

Videomaking, in this sense, is a kind of first step in critical thinking—visually and aurally playing out one's acquired perspective in one's own way. In terms of critical thinking, making a film is a point at which a student can cross the line between learning and living—applying learning to their own expressions.

Inventive Video

What the K–12 teachers and I were finding over time was that our videomaking activities were instilling a very similar spirit to curricula as inventive spelling. Our version, call it *inventive video,* was unlocking "the moviemaker" within young learners in a similar way that inventive spelling was unlocking "the writer" within them. Both of these expressive tools were exercising critical thinking as well.

I had seen this inventive quality of video at many different levels of K–12 and it took on different forms. One of its earliest forms was something I referred to as "the repeat factor."

The Repeat Factor

The repeat factor had to do with the nearly universal desire for young and old (but *especially* young) videomakers to watch what they made over and over and over. One of the great advantages of video as a medium of learning is that it can be repeated. Programs can be watched over and over, and children usually enjoy this. The advantage of repeating is at least twofold—first for reinforcement of instructional or experiential content, and second for critical insight.

Any lesson that a videomaking experience might contain is reinforced (like any learning activity) through repetition. The novelty and entertainment qualities that go with video, not to mention the self-made aspects of it, promote a desire for repetition. This factor is not unlike that of video games. It is repetition that helps children attain exceptional videogame skills. Repetition is a sound educational technique, especially when motivated by the learners. Some of our K–12 videos were about complex subjects (for instance, *Once Upon a Coldheart* and *The Monster Threat*) and needed to be viewed more than once to be truly understood and appreciated.

Many times we would bring a video camera on a field trip, such as a zoo, and "re-experience" the field trip on the screen once we got back to the classroom. This gave the children a chance to study the experience more reflectively and celebrate the experience of doing something together as a class. Any time we watched a field trip—a completely unplanned and unstructured video work as compared to other videos—students would watch it over and over and over. There was never enough time to watch it as many times as they wished to.

The other aspect of the repeat factor was related to critical thinking. In the same way that students talk about how much more they see in a film when they watch it more than once, repeat viewers are lured into critical reflections of repeated viewing experiences. They see and understand more than they originally did. And they look at the experience more critically: looking *into* the content more than simply looking *at* it.

The experience of watching Neddie Marks introduce and narrate *Pollution Rangers* was an example of this. Students, teachers, and parents all commented on his exemplary behavior, which was a far cry from what he was known for. Many, especially after watching his charming performance more than once, actually reconsidered the label they had given him as a classic "difficult student."

There was a similar reaction in Mrs. Beechwood's first-grade class when the students saw their autistic classmate, Dougie, appear amidst them in their video as "normal" as they were. There was a critical level of realization that he truly was no different from them.

Usage of the repeat factor to repeat events and ceremonies—for instance, doing a scene about "the last day of school"—provided an opportunity of enhanced significance of such events in the school experience. Repeated viewing appeared to solidify the significance of such events in a student's memory.

Applied Learning

In its simplest form, inventive video was the experience of young learners applying class concepts on their own. Since most teachers I worked with over the years had no moviemaking experience, they had no preset designs on what videos in their classes should consist of, or even how they should be done. This gave K–12 learners uncharacteristically wide latitudes in doing their video projects.

I recently discovered a sixth-grade video my son had made that I had never seen. Although it was largely a playful video spoof of Greek gods, it was a very clever story and contained a great deal of information about Greek mythology, and it was a critical application of curricula in a child's preferred context: comedy. In the end, my son and his project team took ownership of the lesson.

Problem Solving and Innovation

Videomaking often put young learners in the position of confronting problems and challenges that had no correct (or existing) answers. They needed to think critically to meet the challenge or solve the problem. Diana Green's *Pollution Rangers* movies demonstrated this on many different levels

In the first movie, students faced the challenge of writing a movie that demonstrated how kids could make a difference in the world. This was a contest administered by a party outside their school so they had to interpret on their own just what constituted *making a difference.* Ms. Green discouraged "Pollyanna" notions of doing good and instead encouraged kids to think about things they could actually do.

In the end, they decided on two things: action by example and using their heads. In their "action by example" solution, they wanted to show the importance of simple things: things like not letting water run wastefully when they brushed their teeth, and picking up all litter—even if they weren't responsible for it. The principle behind this was that everything they did—even as little individuals—mattered greatly when it came to taking care of their world. They incorporated this into their character's actions.

In the "using their heads" solution, Ms. Green encouraged her students to contemplate a television story that didn't resort to violence to solve a problem. Much of the media that they watched, including the *Power Ranger* series that their *Pollution Rangers* were roughly based on, resorted to violent acts of some sort in most plots and solutions. They responded to this challenge by having the boy Rangers propose a violent "Power Ranger–style" effort to stop the pollution zombies that would ultimately fail. The girls would advocate for a nonviolent solution based on careful, clever thought. This solution would help the characters and require no superhuman powers on their part.

In the sequel to *Pollution Rangers*, the new Rangers met the old Rangers (six years their senior) and dealt with issues of time and maturity. They also dealt with issues of justice and forgiveness by scripting an ending that rescued and reformed the evil character in their plot.

These second-graders were engaged in critical thinking in the midst of solving story problems. In this sense, their videomaking was a test ground for the application of their own ideas and perspectives, a chance to activate learning and permanently affect students through the experience.

More Than a Class, More Than a Video

> By using video to investigate and address issues in a broader social context, students can become active in their communities as a media advocacy strategy. This is a method that has true potential to correct inaccurate media stereotypes and to call attention to the need for solutions for explicit social problems.
>
> —Kathleen Tyner (1998)

Though no one in Mann's class ever read a word of Tyner's book, they acted as if they were following it to a tee. They used their video camera to investigate and address issues of law and in the process they became more active in their community than they ever would have dreamed.

They named the video *I Am Somebody* and framed it around the arrest of twenty-six young men in their community and their subsequent prosecution based on the RICO Act. Most, if not all, in the class knew at least one of the twenty-six, and many were related to them.

They studied the principles of law that related to the case and they even brought in the federal prosecutor to discuss the case. Despite the fact that the students felt great empathy for the imprisoned young men and the frightening case against them (there were threats of life sentences), they did their best to capture an open-minded discussion of issues around the case without falling victim to their emotions and attachments to the boys.

Next, they scripted a documentary that included:

- A production company logo (HOOD-tv) that represented their bond as a class of concerned citizens
- An introduction with objective: "To shed light on twenty-six men being prosecuted with the RICO Act"
- A dedication sequence in memory of recently murdered area youths
- Bios of the twenty-six men
- Shots of the south-side neighborhood and other city landmarks
- Perspectives on the twenty-six from teachers, coaches, and others who knew them
- A legal view of the RICO law
- Family viewpoints on the case
- A sample of neighborhood viewpoints on the case
- A collage of the families the twenty-six men have left behind

They had hoped to interview the twenty-six young men, but when that fell through because of the sensitivity of the case, they interviewed many of their families. It took the better part of six months to collect all the footage. Although some of the production was done during Mann's class time, most of it was done after school. There was even a speaking engagement at a local talk radio show. The kids who went were beaming after the event. Even before the film was complete, they had piqued the curiosity of their community. By the time the students came to the university to edit their story together, many people in the community had already "reserved" copies of the unfinished film.

One of the most striking aspects of the film, aside from its courageous subject matter, was the way it generated neighborhood pride. Watchers reveled in seeing the shots of their neighborhood, and hearing the voices and opinions of people in their community. The movie may have been focused on the tenuous fate of twenty-six young men, but it was also the story of a tragic set of circumstances all too familiar to poor neighborhoods in American cities.

The film was screened to more than a hundred attendees at a special university screening. The kids introduced the film and it was heralded as a great success. It wouldn't be an exaggeration to say that the video Mann's class made has achieved near mythical status. Part of it has to do with the fact that it looked at the problems of this south-side community square in the eye, and presented them in a public conversation.

And it wasn't easy, by any means. They had to deal with conflicts between families of the imprisoned. They had to be sensitive to legal proceedings and often steer clear of issues that might affect the accused. They had to deal with threats from police who questioned their motives as filmmakers. The officers tried to disrupt some of their shooting and screening activities, and sought involvement in the editorial process before the film was shown to a large group of people. For all the students knew, they could have been charged with conspiracy for making a film that was sympathetic to the boys being charged under RICO.

In the end, most of the twenty-six were convicted of association with gang activity and given lengthy sentences, most in the area of fifteen-plus years in prison. This might be considered a failure if the purpose of the video was only to call attention to the plight of the twenty-six young men. But the video was about a lot more than twenty-six boys. It was about their families and their communities and their city and the hopelessness that surrounded them—but it wasn't a white flag. It was a message of hope and a call to duty for members of the community to rise up and take care of themselves and their community.

The video project for this Business Law class had become real life, and an unforgettable educational experience. Perhaps more important than that, their efforts had actually touched the community. Less than a year after the completion of the movie, four mothers of the imprisoned boys got together to carry the message of the video one step further. They found the courage to approach the federal prosecutor who had sent their boys to prison and work with him to spread a message of hope, healing, and forgiveness to area high schools.

At the end of the assembly, students had the opportunity to ask the mothers and the prosecutor questions. One young man stood up and asked the mothers, "What can I do if everyone I know gives up on me, doesn't believe in me any more?" The mothers didn't hesitate. They walked down the aisle to the young man and embraced him lovingly.

The message that Mann's students had conveyed was that this community had to take care of itself. That message had been heard.

Funny thing, when I talk about the joys of filmmaking to my students, I refer to that ultimate dream we all share in touching the hearts and souls of countless others as something in the reasonably distant future. These high school students had accomplished this right out of the gate—with their first movie. It just goes to show what can be accomplished with a camera, a purpose, and an inventive dose of critical thinking.

Epilogue

Absolute newness tends not to be a defining feature of revolutions. Most (if not all) revolutions exhibit features of continuity so that even the most radical ideas in science prove again and again to be mere transformations of existing traditional ideas.... This is so characteristic a feature of science that some scientists, such as Albert Einstein, have ended up by conceiving their own work to exhibit evolution rather than revolution; a radical transformation and restructuring of what is known or believed rather than the invention of something new.

—I. B. Cohen (1985) on the nature of revolutions

TELEVISION as an instrument of education is such a revolution. Edward R. Murrow "invented" the possibility almost fifty years ago, when he suggested television could "illuminate and even inspire to the extent that humans are determined to use it to those ends" (1958).

Based on the findings from fourteen years of videomaking in K–12 classes, I find that Murrow was absolutely right. The deciding factor involved transforming television from a *read-only* instrument of learning, to a *read and write* instrument of learning. This led to an amazingly simple discovery: Making television is good for children. It's good because it

- connects solidly with the visual world children have grown up in
- reflects their thought process more effectively than letters and words
- directly taps into the yet unrealized reading resource of television
- provides a valuable instrument for increasingly demanding learning environments
- exercises expression and participation in ideas
- encourages and instills collaborative spirit

- gives dimension to lesson plans
- prompts more active involvement in subjects
- extends the playing field of learning
- offers rich contexts to frame, store, and apply knowledge
- makes subjects more relevant to young learners
- challenges children to be inventive and imaginative
- inspires and exercises critical thinking
- instills lifelong learning

Bad News—Good News

However remarkable making television might be, it comes with growing pains. The challenge ahead relates to what Murrow referred to as "extent," in particular "the extent that humans are determined to use" television as a learning tool.

Here lie two challenges that must be overcome in order to realize the educational potential in television. First, educators must put unnecessary fear and skepticism of TV aside, and allow the strength of visual media to align with the learning process. Second, educators must accommodate the technological turbulence that comes with visual media. "Old teachers" must be eased into sometimes intimidating "new tricks." This transformation is not unlike the introduction of computers to school environments: high learning curve with high payoff.

But beyond these initial challenges, as in any good investment, there are long-term rewards ahead, the greatest of which has nothing to do with media or technology. The simple truth is that making television is not a magic bullet when it comes to learning. It simply enables and strengthens the two most important ingredients to learning: student and teacher. This was never more evident than the day I met Ms. Brown's fourth-grade class.

It had been a long day at Corrigan. We had just finished a time-consuming scene in a second-grade class and much more footage would have to be shot before we could finish editing the movie.

As I was packing up the camera gear, fourth-grade assistant teacher Morgan O'Malley came up to me and asked if I might have a minute to talk. The teacher she worked with, Grace Brown, had just been diagnosed with cancer and would be out of the classroom indefinitely. Ms. O'Malley would be running the class in the meantime.

Once the children were told about the situation, they came up with an idea. They had seen the second-graders and me in the hallway making our latest video, and wondered if they might be able to do something of the same for Ms. Brown—along the lines of a *get well* message.

Inside, I was exhausted. It never ceased to amaze me how K–12 teachers could get through a whole school day. I had difficulty getting through two hours! Not only that, but this teacher, and these well-meaning kids in her class, had no idea how much work it was to make a movie, not to mention how time consuming. And I was up to my ears with the second-grade movie.

Something about Ms. O'Malley's eyes—maybe the slight look of desperation, or perhaps the prospect of letting her kids down—struck me. So I picked up the camera gear and said, "Why don't we see what they want to do?" We did the whole thing that afternoon. It was the right thing to do, and in the end, it was easy. As always, the kids figured it out themselves.

<u>FADE UP FROM BLACK TO AN EMPTY SCHOOL HALLWAY. THE CAMERA TURNS LEFT AND JUST AS WE SEE THE DOOR TO GRACE BROWN'S FOURTH-GRADE CLASSROOM, IT SLOWLY OPENS.</u>

WHEN THE CHILDREN IN THE ROOM SEE THE CAMERA, THEY IMMEDIATELY CROWD AROUND THE LENS WITH EXCITED GREETINGS. THE CHILDREN ARE TALKING OVER EACH OTHER BUT THEIR MESSAGE IS CONSISTENT AND CLEAR:

> GIRL
> (WAVING)
> Hi, Ms. Brown!

A BOY STANDS IN FRONT OF HER

> BOY
> We miss you!

ANOTHER BOY PULLS THE CAMERA TO THE RIGHT

> BOY 2
> Please get better!

THE CAMERA PROCEEDS THROUGH THE ROOM, PANNING LEFT AND RIGHT, MAKING SURE TO CAPTURE EVERY CHILD IN THE ROOM. THE BOYS AND GIRLS ARE CLAMORING OVER EACH OTHER TO SHOW THE CAMERA DIFFERENT THINGS—FROM PAPERS TO CLAY FIGURES TO DRAWINGS AND MISCELLANEOUS TRINKETS.

A FOURTH-GRADE GIRL COMES UP TO THE CAMERA AND ASKS IF SHE CAN HOLD IT NOW. THE CAMERA QUICKLY CHANGES POSITIONS, NOW NOTICEABLY SHORTER, BUT CLEARLY THE PERSPECTIVE OF A CHILD.

> THE CHILDREN TAKE TURNS WITH THE CAMERA, SHOW THEIR TEACHER EVERYTHING FROM THEIR CLASSMATES' ANTICS TO THE CONTENTS OF THEIR DESK.
>
> THERE IS NO RHYME OR REASON TO THE PRESENTATION. IT IS GOVERNED ONLY BY THE EXCITEMENT OF CHILDREN WHO MISS THEIR TEACHER. THEY WANT TO SHOW HER WHAT THEY'VE BEEN DOING SINCE SHE'S BEEN GONE.
>
> FADE TO BLACK.

I was told that the movie had made its way to their sick teacher, but I was too busy with the second-grade movie to follow up on it. It seemed to really make the kids happy. Although this video was not connected to a lesson plan per se, it didn't feel out of place. Part of this had to do with the fact that not all lessons in life are out of a curricular plan book. The other part of this experience was that the spirit of videomaking was rooted in something beyond education—something deeply human. The closest comparison I can make is in music: "Who is there that, in logical words, can express the effect music has on us? A kind of inarticulate unfathomable speech, which leads us to the edge of the Infinite, and lets us for moments gaze into that!" (Thomas Carlyle, on heroes, hero-worship and the heroic in history [1859]).

Moving-image media have a similar, almost spiritual effect that allows souls to share experiences outside of the confines of individual existences. Filmmakers delight in audiences sharing their messages. In a lot of ways, teachers are very similar to filmmakers: They both want to give. Maybe that's why television and film graduates—those who don't end up in the business—are attracted to careers in teaching. They both share an immense sense of giving. In a lot of ways, making this one-day movie made me feel like I was closing in on a deeper sense of what it means to be human and sharing that experience with a very determined bunch of fourth-graders.

Later that summer, I found out at least one more person had shared the same feeling about the experience.

July 15

Dear Mr. Schoonmaker,

At the end of the year we always make a list of people who have helped us during the year so that we can thank them.

When we did that this year, you were right up there on the top of the list! The kids didn't know your name but they clearly knew they wanted to thank "the man who helped us make the video for you."

When I asked them if they had sent you a "thank you," they couldn't remember. So they wanted to make you a certificate . . . and be sure that you know they said "thanks."

Kids are great, aren't they?

And so are the wonderful parents we have at Corrigan Elementary.

I wanted to add my thanks, too, even though it is long overdue. I appreciated so much seeing those kids when I was sick. They really gave me the strength to get back in there again. Thank you for your time and for your patience and understanding with them.

I hope you will have a great summer.

Gratefully,
Grace Brown

Yes, making television was good for the children in Ms. Brown's class, and it also happened to be especially good for Ms. Brown too, which goes to show an even wider potential for the application of videomaking.

But I know Ms. Brown would want to point out that if we get it right with children, everything else should fall into place just fine.

REFERENCES

Adams, Marilyn Jager. 1990. *Beginning to Read: Thinking and Learning about Print.* Urbana-Champaign: University of Illinois, Reading Research and Education Center.
Addison, Joseph. 1710. *Tatler,* no. 147 (18 March).
Allard, H. 1977. *Miss Nelson Is Missing.* Illustrated by James Marshall. Boston: Houghton Mifflin.
Allen, R. 1985. *Speaking of Soap Operas.* Chapel Hill: University of North Carolina Press.
Anderson, D. R., L. F. Alwitt, E. P. Lorch, and S. R. Levin. 1979. Watching Children Watch Television. In *Attention and Cognitive Development,* edited by G. Hale and M. Lewis, 331–62. New York: Plenum.
Barnhurst, K. 1987. *The Knowing Eye: An Applied Arts Approach to Visual Knowledge.* Paper presented at the 1987 AEJMC convention, San Antonio, Texas.
Bazalgette, Cary, and David Buckingham (Eds.). 1995. *In Front of the Children: Screen Entertainment and Young Audiences.* London: British Film Institute.
Berlo, D. 1975. The Context for Communication. In *Communication and Behavior,* edited by G. Hanneman and W. McEwen. Reading, MA: Addison-Wesley.
Bettelheim, B. 1979. Violence: A Neglected Mode of Behavior. In *Surviving and Other Essays.* New York: Alfred A. Knopf.
Bianculli, D. 2000. *Taking Television Seriously.* Syracuse, NY: Syracuse University Press.
Blumer, H. 1969. *Symbolic Interactionism: Perspective and Method.* Englewood Cliffs, NJ: Prentice-Hall.
———. 1990. *Industrialization as an Agent of Social Change.* New York: A. de Gruyter.
Brown, J. A. 1991. *Television Critical Viewing Skills Education: Major Media Literacy Projects in the United States and Selected Countries.* Hinsdale, NJ: Lawrence Erlbaum.
Bruner, J., and S. Weisser. 1991. The Invention of Self: Autobiography and Its Forms. In *Literacy and Orality,* edited by D. R. Olson and N. Torrance. Cambridge, England: Cambridge University Press.
Buckingham, D. 1993a. *Children Talking Television: The Making of Television Literacy.* Basingstoke, England: Falmer Press.
———. 1993b. *Reading Audiences: Young People and the Media.* Manchester, England: Manchester University Press.
———. 2000. *After the Death of Childhood: Growing Up in the Age of Electronic Media.* Cambridge, England: Polity Press.

———. 2002. Introduction: The Child and the Screen. In *Small Screens: Television for Children*. London: Leicester University Press.
Buckingham, David, Hannah Davies, Ken Jones, and Peter Kelley. 1999. *Children's Television in Britain: History, Discourse and Policy*. London: British Film Institute.
Business Coalition for Education Reform. 2001. *Why Business Cares about Education*, at www.bcer.org.
Carlsson-Paige, N., and D. Levin. 1990. *Who's Calling the Shots*. Philadelphia: New Society.
Carlyle, T. 1859. *On Heroes, Hero-Worship, and the Heroic in History: Six Lectures*. New York: John Wiley.
Clark, R., and G. Salomon. 1986. Media in Teaching. In *Handbook of Research on Teaching*, 3rd ed., edited by M. C. Wittrock, 464–78. New York: Macmillan.
Cohen, I. B. 1985. *Revolution in Science*. Cambridge, MA: Harvard University Press.
Critical Thinking Community. 1999. Foundation and Center for Critical Thinking, 16 April, at www.criticalthinking.org.
Critical Thinking: What It Is and Why It Counts. 2004. California Academic Press, at www.criticalthinking.org (accessed 2 April 2005).
Cullingford, C. 1984. *Children and Television*. New York: St. Martin's Press.
Dahlgren, P. 1995. *Television and the Public Sphere: Citizenship, Democracy and the Media*. London: Sage.
Davidson, Hall. 2004. *Meaningful Digital Video for Every Classroom*. TechLearning, 5 April, at http://techlearning.com/story/showArticle.jhtml?articleID=18700330 (accessed 24 November 2004).
Denzin, N. 1995. *The Cinematic Society*. London: Sage.
Emme, M. 1995. *Life, Death and Invisibility*. Media-L, 16 August, at Media-L@nmsu.edu (accessed 15 February 1996).
Encarta World English Dictionary. 1999. Microsoft Corporation. Developed for Microsoft by Bloomsbury. CD-ROM.
Engelhardt, T. 1987. The Shortcake Strategy. In *Watching Television: A Pantheon Guide to Popular Culture*, edited by T. Gitlin, 68–110. New York: Pantheon.
Ewen, S., and E. Ewen. 1982. *Channels of Desire: Mass Images and the Shaping of American Consciousness*. New York: McGraw Hill.
Fish, S. 1980. *Is There a Text in This Class? The Authority of Interpretive Communities*. Cambridge, MA: Harvard University Press.
Fowlie, J. 1995. *Turn off TV: A Media Literacy Listserve Exchange*, 30 October, at Media-l @nmsu.edu.
Fulgham, Robert. 2003. *All I Really Need to Know I Learned in Kindergarten*. New York: Ballantine Books.
Gardner, H. 1983. *Frames of Mind: The Theory of Multiple Intelligences*. New York: Basic Books.
———. 1985. *Frames of Mind*. New York: Basic Books.
Gauntlett, D. 1995. *Moving Experiences: Understanding Television's Influences and Effects*. London: John Libbey.
———. 1996. *Video Critical: Children, the Environment and Media Power*. Luton: John Libbey.
Gitlin, T. 1987. *Watching Television*. New York: Pantheon Books.

Gordon, Edwin E. 1990. *A Music Learning Theory for Newborn and Young Children.* Chicago: GIA Publications.

Greenberg, B., J. Brown, and N. Buerkel-Rothfuss. 1993. *Media, Sex and the Adolescent.* Cresskill, NJ: Hampton Press.

Gross, L. 1974. Symbolic Competence. In *Media and Symbols: The Forms of Expression, Communication, and Education,* edited by D. R. Olson. Chicago: National Society for the Study of Education.

Hefzallah, I. 1987. *Critical Viewing of Television: A Book for Parents and Teachers.* Lanham, MD: University Press of America.

———. 1990. *The New Learning and Telecommunications Technologies: Their Potential Applications in Education.* Springfield, IL: Charles C. Thomas.

Hobbs, R. 1998. The Seven Great Debates in the Media Literacy Movement. *Journal of Communication* 48(1): 16–32.

Hodge, B., and D. Tripp. 1986. *Children and Television: A Semiotic Approach.* Cambridge, England: Polity Press.

Hooton, Earnest. 1991. *International Herald Tribune,* May 2.

Horton, R., and R. Finnegan (Eds.). 1973. *Modes of Thought.* London: Faber.

Hutchinson, T. 1946. *Here Is Television: Your Window to the World.* New York: Dial Press.

Jensen, E., D. Osborne, A. Pogrebin, and T. Rose. 1998. Consumer Alert. *Brill's Content,* October, 130–42.

Katz, Jon. 1997. *Virtuous Reality: How America Surrendered Discussion of Moral Values to Opportunists, Nitwits and Blockheads Like William Bennett.* New York: Random House.

Kipping, P. L. 1995a. *Horrors of Horrors.* Media-L, 16 August, at Media-L@nmsu.edu (accessed 17 February 1996).

———. 1995b. *Life, Death and Invisibility — Reply.* Media-L, 17 August, at Media-L@nmsu.edu (accessed 17 February 1996).

Lasswell, H. 1948. The Structure and Function of Communication in Society. In *The Communication of Ideas,* edited by L. Bryson. New York: Harper and Bros.

Liebes, T., and E. Katz. 1990. *The Export of Meaning: Cross-Cultural Readings of Dallas.* Oxford: Oxford University Press.

Lloyd-Kolkin, Donna, and Kathleen R. Tyner. 1991. *Media and You: An Elementary Media Literacy Curriculum.* Englewood Cliffs, NJ: Educational Technology.

Lubbock, Sir John. 1889. *The Pleasures of Life.* London: Macmillan and Co. Presented by Authorama: Public Domain Books, at www.authorama.com/pleasures-of-life-13.html.

Luke, C. 1990. *Constructing the Child Viewer: A History of the American Discourse on Television and Children, 1950–1980.* New York: Praeger.

Marshall, James. 2002. *Learning with Technology: Evidence That Technology Can, and Does Support Learning.* Prepared in May 2002 for Cable in the Classroom, at http://medialit.org/reading_room/article545.html.

Marvin, C. 1988. *When Old Technologies Were New.* New York: Oxford University Press.

Maxwell, William. 1980. *So Long, See You Tomorrow.* New York: Knopf.

McCloud, S. 1993. *Understanding Comics: The Invisible Art.* New York: Harper Collins.

McLuhan, M. 1964. *Understanding Media: The Extensions of Man.* New York: McGraw-Hill.

McQuail, D. 1994. *Mass Communication Theory.* London: Sage.

Messaris, Paul. 1994. *Visual Literacy: Image, Mind, and Reality*. Boulder, CO: Westview Press.
Meyer, L. B. 1956. *Emotion and Meaning in Music*. Chicago: University of Chicago Press, 40–41.
Moores, S. 1993. *Interpreting Audiences: The Ethnography of Media Consumption*. London: Sage.
Murrow, Edward. 1958. RTNDA Convention, Chicago, October 15, *Radio-Television News Directors Association*, at www.rtnda.org/resources/speeches/murrow.shtml.
Neuman, S. B. 1991. *Literacy in the Television Age*. Norwood, NJ: Ablex.
Newcomb, H. 1974. *TV: The Most Popular Art*. Garden City, NY: Anchor Books.
Newman, B., and J. Mara. 1995. *Reading, Writing, and TV: A Video Handbook for Teachers*. Fort Atkinson, WI: Highsmith Press.
Nickerson, R. 1993. On the Distribution of Cognition. In *Distributed Cognitions: Psychological and Educational Considerations*, edited by G. Salomon. New York: Cambridge University Press.
Norton, Donna. 1983. *Through the Eyes of a Child*. Columbus: Charles E. Merrill.
Olson, D. R. (Ed.). 1974. *Media and Symbols: The Forms of Expression, Communication, and Education*. Chicago: National Society for the Study of Education.
———. 1977. From Utterance to Text: The Bias of Language in Speech and Writing. *Harvard Educational Review* 47:257–81.
———. 1987. Schooling and the Transformation of Common Sense. In *Common Sense: The Foundations for Social Science*, edited by F. Van Holthoon and D. R. Olson, 319–40. Lanham, MD: University Press of America.
———. 1994. *The World on Paper: The Conceptual and Cognitive Implications of Writing and Reading*. Cambridge, England: Cambridge University Press.
Olson, D. R., A. Hildyard, and N. Torrance. 1985. *Literacy, Language and Learning: The Nature and Consequences of Reading and Writing*. New York: Cambridge University Press.
Olson, D. R., and N. Torrance (Eds.). 1991. *Literacy and Orality*. Cambridge, England: Cambridge University Press.
Palmer, E. L. 1993. *Toward a Literate World: Television in Literacy Education—Lessons from the Arab Region*. Boulder, CO: Westview Press.
Palmer, Patricia. 1986. *The Lively Audience: A Study of Children around the TV Set*. Sydney: Allen and Unwin.
Paul, R., A. J. A. Binker, K. Adamson, and D. Martin. 1989. *Critical Thinking Handbook: High School—A Guide for Redesigning Instruction*. Rohnert Park, CA: Center for Critical Thinking and Moral Criticism, Sonoma State University.
Phillips, P. 1991. *Saturday Morning Mind Control*. Nashville, TN: Oliver-Nelson Books.
Postman, N. 1979. The First Curriculum: Comparing School and Television. *Phi Delta Kappan* (November): 163–71.
———. 1985. *Amusing Ourselves to Death*. New York: Penguin Books.
Potter, W. J. 1998. *Media Literacy*. Thousand Oaks, CA: Sage.
Prensky, Marc. 2001. *Digital Game-Based Learning*. New York: McGraw-Hill.
Provost, Gary. 1990. *Make Your Words Work*. Cincinnati, OH: Writer's Digest Books.
Rabiger, Michael. 2003. *Directing: Film Techniques and Aesthetics*. London: Focal Press.

Reeves, B. 1978. Perceived TV Reality as a Predictor of Children's Social Behavior. *Journalism Quarterly* 55:682–89, 695.

Rowland, W. 1983. *The Politics of TV Violence: Policy Uses of Communication Research.* Beverly Hills: Sage.

Salomon, G. 1982. Television Literacy and Television vs. Literacy. *Journal of Visual and Verbal Languaging* 2(2): 7–17.

——— (Ed.). 1993. *Distributed Cognitions: Psychological and Educational Considerations.* New York: Cambridge University Press.

Sandved, Kjell, and M. Emsley. 1975. *Butterfly Magic.* New York: Viking Press.

Schoonmaker, M. 1994. The Character Viewing in a Video Community. PhD diss., Syracuse University.

Sefton-Green, Julian (Ed.). 1999. *Young People, Creativity and New Technologies: The Challenge of Digital Arts.* London: Routledge.

Selfe, Cynthia L. 1999. *Technology and Literacy in the Twenty-First Century: The Importance of Paying Attention.* Carbondale: Southern Illinois University Press.

Sessions, R. 1968. *The Musical Experience.* New York: Athenaeum.

Shaw, D. 2003. A Plea for Media Literacy in Our Nation's Schools. *Los Angeles Times,* November 30, at www.medialit.org/reading_room/article631.html.

Siepman, C. 1962. The Missing Literature of Television. In *The Eighth Art,* edited by R. L. Shayon, 224–25. New York: Holt, Rinehart & Winston.

Silverstone, R. 1994. *Television and Everyday Life.* London: Routledge.

Smith, D. 1991. *Video Communication: Structuring Content for Maximum Program Effectiveness.* Belmont, CA: Wadsworth.

Sontag, S. 1966. *Against Interpretation.* New York: Farrar, Strauss & Giroux.

Strauss, A. 1987. *Qualitative Analysis for Social Scientists.* Cambridge, England: Cambridge University Press.

Street, B. V. 1984. *Literacy in Theory and Practice.* Cambridge, England: Cambridge University Press.

Suss, Daniel. 2001. Computers and the Internet in School: Closing the Knowledge Gap? In *Children and Their Changing Media Environment: A European Comparative Study,* edited by Sonia Livingstone and Moira Bovill. Mahwah, NJ: Lawrence Erlbaum.

Sylwester, R. 1995. *A Celebration of Neurons: An Educator's Guide to the Human Brain.* Alexandria, VA: Association for Supervision and Curriculum Development.

Sylwester, R., and H. Marcinkiewicz. 2003. The Brain, Technology, and Education: An Interview with Robert Sylwester. The Technology Source (November/December), at http://technologysource.org/article/brain_technology_and_education.

Thoman, E., and T. Jolls. 2004. A National Priority for a Changing World. *American Behavioral Scientist* 48 (1), (September): 18–29.

Timmerman, J. 2002. *Jane Kenyon: A Literary Life.* Grand Rapids, MI: William B. Eerdmans Publishing Co.

Twitchell, J. 1989. *Preposterous Violence: Fables of Aggression in Modern Culture.* New York: Oxford University Press.

Tyner, Kathleen. 1994. Video in the Classroom: A Tool for Reform. *Arts Education Policy Review* 96(1), (September/October).

———. 1995. *Re: Jo Holz presentation*, 29 September, at media-l@nmsu.edu (accessed 29 September 1995).

———. 1998. *Literacy in a Digital World: Teaching and Learning in the Age of Information*. Mahwah, NJ: Lawrence Erlbaum Associates.

Using Inventive Spelling. 1999. LinguaLinks Library, Version 4.0, CD-ROM. 16 March 1999, from SIL International, at www.sil.org/lingualinks/literacy/ImplementALiteracyProgram/UsingInventiveSpelling.htm (accessed 9 April 2005).

VanMaanen, John. 1988. *Tales of the Field: On Writing Ethnography*. Chicago: University of Chicago Press.

VSDA. 1999. *Video Software Dealers Association*, June, at www.vsda.org.

West, M. 1997. *Trust Your Children: Voices against Censorship in Children's Literature*, 2nd ed. New York: Neal-Schuman.

Willis, Paul, et al. 1990. *Common Culture: Symbolic Work at Play in the Everyday Cultures of the Young*. Boulder, CO: Westview Press.

Zettl, H. 1990. *Sight, Sound Motion: Applied Media Aesthetics*, 2nd ed. Belmont, CA: Wadsworth.

Photo Credits

Introduction: Video still-frame from *Invisible Juice* by Mrs. Brighton's kindergarten class.

Chapter 1: Video still-frame from *Pollution Rangers II* by Ms. Green's second-grade class.

Chapter 2: Video still-frame from *Imag-A-Book and the Read-A-Lot Gang* by Mr. Samson's high school video-production class.

Chapter 3: Video still-frame from *The Culture Project* by Mrs. Spencer's third-grade class.

Chapter 4: Video still-frame from *The Monster Threat* by Mrs. Redfield's third-grade class.

Chapter 5: Photo by Tessa K. Ferrario.

Chapter 6: Video still-frame from *The Mrs. Beechwood Show* by Mrs. Beechwood's first-grade class.

Chapter 7: Video still-frame from *News Class 104* by Ms. Bronson's fourth-grade class.

Chapter 8: Video still-frame from *The Pollution Rangers* by Ms. Green's second-grade class.

Chapter 9: Video still-frame from *The Poetry Virus* by Mrs. Catrell's fifth-grade class.

Chapter 10: Video still-frame from *Once Upon a Coldheart* by Mrs. Brighton's kindergarten class.

Chapter 11: Video still-frame from *I Am Somebody* by Mr. Mann's high school business law class.